# ARTIFICIAL NEURAL NETWORKS FOR MODELLING AND CONTROL OF NON-LINEAR SYSTEMS

# Artificial Neural Networks for Modelling and Control of Non-Linear Systems

by

Johan A. K. Suykens
Joos P. L. Vandewalle
Bart L. R. De Moor

KLUWER ACADEMIC PUBLISHERS
BOSTON / DORDRECHT / LONDON

A C.I.P. Catalogue record for this book is available from the Library of Congress

ISBN 978-1-4419-5158-8

---

Published by Kluwer Academic Publishers,
P.O. Box 17, 3300 AA Dordrecht, The Netherlands.

Kluwer Academic Publishers incorporates
the publishing programmes of
D. Reidel, Martinus Nijhoff, Dr W. Junk and MTP Press.

Sold and distributed in the U.S.A. and Canada
by Kluwer Academic Publishers,
101 Philip Drive, Norwell, MA 02061, U.S.A.

In all other countries, sold and distributed
by Kluwer Academic Publishers Group,
P.O. Box 322, 3300 AH Dordrecht, The Netherlands.

*Printed on acid-free paper*

# Contents

**Preface** ix

**Notation** xi

**1 Introduction** 1
- 1.1 Neural information processing systems . . . . . . . . . . . . . 1
- 1.2 ANNs for modelling and control . . . . . . . . . . . . . 5
- 1.3 Chapter by Chapter overview . . . . . . . . . . . . . 8
- 1.4 Contributions . . . . . . . . . . . . . 15

**2 Artificial neural networks: architectures and learning rules** 19
- 2.1 Basic neural network architectures . . . . . . . . . . . . . 19
- 2.2 Universal approximation theorems . . . . . . . . . . . . . 23
  - 2.2.1 Multilayer perceptrons . . . . . . . . . . . . . 23
  - 2.2.2 Radial basis function networks . . . . . . . . . . . . . 27
- 2.3 Classical paradigms of learning . . . . . . . . . . . . . 28
  - 2.3.1 Backpropagation . . . . . . . . . . . . . 29
  - 2.3.2 RBF networks . . . . . . . . . . . . . 33
- 2.4 Conclusion . . . . . . . . . . . . . 35

**3 Nonlinear system identification using neural networks** 37
- 3.1 From linear to nonlinear dynamical models . . . . . . . . . . . . . 38
- 3.2 Parametrization by ANNs . . . . . . . . . . . . . 39
  - 3.2.1 Input/output models . . . . . . . . . . . . . 39
  - 3.2.2 Neural state space models . . . . . . . . . . . . . 41
  - 3.2.3 Identifiability . . . . . . . . . . . . . 43
- 3.3 Learning algorithms . . . . . . . . . . . . . 45
  - 3.3.1 Feedforward network related models . . . . . . . . . . . . . 46
    - 3.3.1.1 Backpropagation algorithm . . . . . . . . . . . . . 46
    - 3.3.1.2 Prediction error algorithms . . . . . . . . . . . . . 46

3.3.1.3 Extended Kalman filtering . . . . . . . . . . . . 48
3.3.2 Recurrent network related models . . . . . . . . . . . . 50
3.3.2.1 Dynamic backpropagation . . . . . . . . . . . . 50
3.3.2.2 Extended Kalman filtering . . . . . . . . . . . . 54
3.4 Elements from nonlinear optimization theory . . . . . . . . . . . 55
3.5 Aspects of model validation, pruning and regularization . . . . 58
3.6 Neural network models as uncertain linear systems . . . . . . . 61
3.6.1 Convex polytope . . . . . . . . . . . . . . . . . . . 62
3.6.2 LFT representation . . . . . . . . . . . . . . . . . . 65
3.7 Examples . . . . . . . . . . . . . . . . . . . . . . . . . . 68
3.7.1 Some challenging examples from the literature . . . . . 68
3.7.2 Simulated nonlinear system with hysteresis . . . . . . . 69
3.7.3 Identification of a glass furnace . . . . . . . . . . . . . 75
3.7.4 Identifying $n$-double scrolls . . . . . . . . . . . . . . . 77
3.8 Conclusion . . . . . . . . . . . . . . . . . . . . . . . . . 82

**4 Neural networks for control** **83**
4.1 Neural control strategies . . . . . . . . . . . . . . . . . . . 83
4.1.1 Direct versus indirect adaptive methods . . . . . . . . . 83
4.1.2 Reinforcement learning . . . . . . . . . . . . . . . . . 85
4.1.3 Neural optimal control . . . . . . . . . . . . . . . . . . 87
4.1.4 Internal model control and model predictive control . . 88
4.2 Neural optimal control . . . . . . . . . . . . . . . . . . . . . 90
4.2.1 The $N$-stage optimal control problem . . . . . . . . . . 90
4.2.2 Neural optimal control: full state information case . . . 92
4.2.3 Stabilization problem: full static state feedback . . . . . 92
4.2.4 Tracking problem: the LISP principle . . . . . . . . . . 94
4.2.5 Dynamic backpropagation . . . . . . . . . . . . . . . . 95
4.2.6 Imposing constraints from linear control theory . . . . . 96
4.2.6.1 Static feedback using feedforward nets . . . . . 97
4.2.6.2 Dynamic feedback using recurrent nets . . . . 99
4.2.6.3 Transition between equilibrium points . . . . . 101
4.2.6.4 Example: swinging up an inverted pendulum . 104
4.2.6.5 Example: swinging up a double inverted pendulum . . . . . . . . . . . . . . . . . . . . . . . . 111
4.3 Conclusion . . . . . . . . . . . . . . . . . . . . . . . . . 115

**5 $NL_q$ Theory** 117
5.1 A neural state space model framework for neural control design 118
5.2 $NL_q$ systems . . . . . 122
5.3 Global asymptotic stability criteria for $NL_q$s . . . . . 127
5.3.1 Stability criteria . . . . . 127
5.3.2 Discrete time Lur'e problem . . . . . 132
5.4 Input/Output properties - $l_2$ theory . . . . . 134
5.4.1 Equivalent representations for $NL_q$s . . . . . 134
5.4.2 Main Theorems . . . . . 136
5.5 Robust performance problem . . . . . 140
5.5.1 Perturbed $NL_q$s . . . . . 140
5.5.2 Connections with $\mu$ theory . . . . . 145
5.6 Stability analysis: formulation as LMI problems . . . . . 147
5.7 Neural control design . . . . . 150
5.7.1 Synthesis problem . . . . . 151
5.7.2 Non-convex nondifferentiable optimization . . . . . 152
5.7.3 A modified dynamic backpropagation algorithm . . . . . 153
5.8 Control design: some case studies . . . . . 154
5.8.1 A tracking example on diagonal scaling . . . . . 154
5.8.2 A collection of stabilization problems . . . . . 157
5.8.3 Mastering chaos . . . . . 162
5.8.4 Controlling nonlinear distortion in loudspeakers . . . . . 164
5.9 $NL_q$s beyond control . . . . . 168
5.9.1 Generalized CNNs as $NL_q$s . . . . . 168
5.9.2 LRGF networks as $NL_q$s . . . . . 172
5.10 Conclusion . . . . . 175

**6 General conclusions and future work** 177

**A Generation of $n$-double scrolls** 181
A.1 A generalization of Chua's circuit . . . . . 182
A.2 $n$-double scrolls . . . . . 185

**B Fokker-Planck Learning Machine for Global Optimization** 195
B.1 Fokker-Planck equation for recursive stochastic algorithms . . . 196
B.2 Parametrization of the pdf by RBF networks . . . . . 198
B.3 FP machine: conceptual algorithm . . . . . 200
B.4 Examples . . . . . 203
B.5 Conclusions . . . . . 205

**C Proof of $NL_q$ Theorems** 207

**Bibliography** 215

**Index** 233

# Preface

The topic of this book is the use of artificial neural networks for modelling and control purposes. The relatively young field of neural control, which started approximately ten years ago with Barto's broomstick balancing experiments, has undergone quite a revolution in recent years. Many methods emerged including optimal control, direct and indirect adaptive control, reinforcement learning, predictive control etc. Also for nonlinear system identification, many neural network models and learning algorithms appeared. Neural network architectures like multilayer perceptrons and radial basis function networks have been used in different model structures and many on- or off-line learning algorithms exist such as static and dynamic backpropagation, prediction error algorithms, extended Kalman filtering, to name a few.

The abundance of methods is basically due to the fact that neural network models and control systems form just another class of nonlinear systems, and can of course be approached from many theoretical points of view. Hence, for newcoming people, interested in this area, and even for experienced researchers it might be hard to get a fast and good overview of the field. The aim of this book is precisely to present both classical and new methods for nonlinear system identification and neural control in a straightforward way, with emphasis on the fundamental concepts.

The book results from the first author's PhD thesis. One major contribution is the so-called '$NL_q$ theory', described in Chapter 5, which serves as a unifying framework for stability analysis and synthesis of nonlinear systems that contain linear and static nonlinear operators that satisfy a sector condition. $NL_q$ systems are described by nonlinear state space equations with $q$ layers and hence encompass most of the currently used feedforward and recurrent neural networks. Using neural state space models, the theory enables to design controllers based upon identified models from measured input/output data. It turns out that many problems arising in neural networks, systems and control can be considered as special cases of $NL_q$ systems. It is also shown by examples how different types of behaviour, ranging from globally asymptotically stable systems, systems with multiple equilibria, periodic and chaotic behaviour can be mastered within $NL_q$ theory.

## Acknowledgements

We thank L. Chua, P. Dewilde, A. Barbé and D. Bollé for participation in the jury of the thesis, from which this book originates. The leading research work of L. Chua on chaos and cellular neural networks formed a continuous source of inspiration for the present work, as one can observe from the $n$-double scroll, generalized CNNs within $NL_q$ theory and the work on identifying and mastering chaotic behaviour. The summer course of S. Boyd at our university on 'convex optimization optimization in control design' in 1992 was the start for studying neural control systems from the viewpoint of (post)modern control theory, finally leading to the development of $NL_q$ theory in this book. At this point we also want to thank L. El Ghaoui, P. Gahinet, P. Moylan, A. Tesi, P. Kennedy, P. Curran, M. Hasler, T. Roska and our colleagues J. David, L. Vandenberghe and C. Yi for stimulating discussions.

Furthermore we thank all our SISTA colleagues for the pleasant atmosphere. Especially we want to mention here the people that have been working on neural networks, D. Thierens, J. Dehaene, Y. Moreau, J. Hao and S. Tan. We are also grateful to all our colleagues from the *Interdisciplinary Center for Neural Networks* of the K.U. Leuven, which is a forum for mathematicians, physicists, engineers and medical researchers to meet each other on a regular basis. From Philips Belgium we thank J. Van Ginderdeuren, C. Verbeke and L. Auwaerts for our common research work on reducing nonlinear distorsion in loudspeakers.

The structure of this book partially originated from a series of lectures for the 'Belgian Graduate school on Systems and Control' on 'Artificial neural networks with application to systems and control' in 1993. The work reported in this book was supported by the Flemish Community through the Concerted Action Project *Applicable Neural Networks*, and the framework of the Belgian Programme on Interuniversity Poles of Attraction, initiated by the Belgian State, Prime Minister's Office for Science, Technology and Culture (IUAP-17 and IUAP-50). We thank all the people that were involved in these frameworks for the many stimulating interactions.

Johan Suykens
Joos Vandewalle
Bart De Moor
K.U. Leuven, Belgium

# Notation

## Symbols

| | |
|---|---|
| $\mathbb{R}^n(\mathbb{C}^n)$ | set of real (complex) column vectors |
| $\mathbb{R}^{n\times m}(\mathbb{C}^{n\times m})$ | set of real (complex) $n \times m$ matrices |
| $a^i_j, a_{ij}$ | $ij$-th entry of matrix $A$ (unless locally overruled) |
| $A^T$ | transpose of matrix $A$ |
| $A^*$ | complex conjugate transpose of $A$ |
| $\|.\|$ | absolute value of a scalar |
| $\|\|x\|\|_2, x \in \mathbb{R}^n$ | 2-norm of vector $x$, $\|\|x\|\|_2 = (\sum_{i=1}^n \|x_i\|^2)^{1/2}$ |
| $\|\|x\|\|_\infty$ | $\infty$-norm of vector $x$, $\|\|x\|\|_\infty = \max_i \|x_i\|$ |
| $\|\|x\|\|_1$ | 1-norm of vector $x$, $\|\|x\|\|_1 = \sum_{i=1}^n \|x_i\|$ |
| $\|\|x\|\|_p$ | $p$-norm vector $x$, $\|\|x\|\|_p = (\sum_{i=1}^n \|x_i\|^p)^{1/p}$ |
| $\|\|A\|\|_2, A \in \mathbb{R}^{n\times m}$ | induced 2-norm of matrix $A$, $\|\|A\|\|_2 = \overline{\sigma}(A)$ |
| $\|\|A\|\|_\infty$ | induced $\infty$-norm of matrix $A$, $\|\|A\|\|_\infty = \max_i \sum_{j=1}^m \|a_{ij}\|$ |
| $\|\|A\|\|_1$ | induced 1-norm of matrix $A$, $\|\|A\|\|_1 = \max_j \sum_{i=1}^n \|a_{ij}\|$ |
| $\sigma_i(A)$ | $i$-th singular value of $A$ |
| $\overline{\sigma}(A)$ | maximal singular value of $A$, $\overline{\sigma}(A) = (\lambda_{max}(A^*A))^{1/2}$ |
| $\underline{\sigma}(A)$ | minimal singular value of $A$ |
| $\kappa(A)$ | condition number of $A$, $\kappa(A) = \|\|A\|\|_2\|\|A^{-1}\|\|_2$ |
| $\lambda_i(A)$ | $i$-th eigenvalue of $A$ |
| $\rho(A)$ | spectral radius of $A$, $\rho(A) = \max_i \|\lambda_i(A)\|$ |
| $A > 0$ | $A$ positive definite |
| $A > B$ | $A - B$ positive definite |
| $[x; y]$ | concatenation of vectors $x$ and $y$ (Matlab notation) |
| $[A\ B; C\ D]$ | matrix consisting of block rows $[A\ B]$ and $[C\ D]$ |
| $A(i, :)$ | $i$-th row of matrix $A$ (Matlab notation) |
| $A(:, j)$ | $j$-th column of matrix $A$ (Matlab notation) |
| $I_n$ | identity matrix of size $n \times n$ |
| $O_{m\times n}$ | zero matrix of size $m \times n$ |
| $l_2^n$ | set of square summable sequences in $\mathbb{C}^n$ |
| $\|\|e\|\|_2, e \in l_2^n$ | $l_2$ norm of a sequence $e$, $\|\|e\|\|_2^2 = \sum_{k=1}^\infty \|\|e_k\|\|_2^2$ |
| $f(.; \theta)$ | function $f$, parametrized by $\theta$ |

# Acronyms

| | |
|---|---|
| ANN | Artificial Neural Network |
| $NL_q$ | multilayer dynamical system with alternating static Nonlinear and Linear operators ($q$ layers) |
| LFT | Linear Fractional Transformation |
| LMI | Linear Matrix Inequality |
| I/O | Input/Output |
| SISO | Single Input Single Output |
| MIMO | Multi Input Multi Output |
| RBF | Radial Basis Function |
| NARX | Nonlinear AutoRegressive with eXternal input |
| NARMAX | Nonlinear AutoRegressive Moving Average with eXternal input |
| EKF | Extended Kalman Filter |
| LQ | Linear Quadratic |
| SQP | Sequential Quadratic Programming |
| GCNN | Generalized Cellular Neural Network |
| LRGF | Locally Recurrent Globally Feedforward |
| ODE | Ordinary Differential Equation |
| PDE | Partial Differential Equation |
| FP | Fokker-Planck |

# $NL_q$ theory: signals and systems

| | |
|---|---|
| $\mathcal{M}_i$ | Neural state space model ($i \in \{0, 1, 2, 3\}$) |
| $\mathcal{C}_j$ | Neural state space controller ($j \in \{0, 1, ..., 5\}$) |
| $\Xi_j^i$ | Family of neural control problems for $\mathcal{M}_i$ and $\mathcal{C}_j$ |
| $\mathcal{S}_i$ | Standard plant model related to model $\mathcal{M}_i$ |
| $k$ | discrete time index |
| $w_k$ | exogenous input |
| $d_k$ | reference input |
| $\epsilon_k$ | white noise input |
| $e_k$ | regulated output |
| $u_k$ | control signal |
| $y_k$ | sensed output |
| $\hat{x}_k$ | state of model $\mathcal{M}_i$ |
| $z_k$ | state of controller $\mathcal{C}_j$ |
| $p_k$ | state of $NL_q$ system for standard plant configuration |

# Chapter 1

# Introduction

In this Introduction we give first a short explanation about neural information processing systems in Section 1.1, including basic architectures, learning modes and some brief history. In Section 1.2 we motivate the use of artificial neural networks for modelling and control. In Section 1.3 we sketch the broad picture of this book, together with a Chapter by Chapter overview. In Section 1.4 own contributions are listed.

## 1.1 Neural information processing systems

Artificial neural networks (ANNs) form a class of systems that is inspired by biological neural networks. They usually consist of a number of simple processing elements, called neurons, that are interconnected to each other. In most cases one or more layers of neurons are considered that are connected in a feedforward or recurrent way (Zurada (1992), Grossberg (1988), Lippmann (1987)). The strength of the interconnections is quantified by means of interconnection weights. Basic features of neural architectures are that they work massively parallel, the weights have to be learned from a set of examples and can be adapted. Although ANNs can perform human brain-like tasks such as object and pattern recognition, speech recognition or associative memory, there is still a huge gap between biological and artificial neural nets. Nevertheless, although we are still far away from mimicing the human brain, from an engineering point of view, it is certainly a fruitful step to let us inspire by biology. Indeed ANNs have provided good solutions to many problems in various fields: examples include classification problems, vision, speech, signal processing, time series prediction, modelling and control, robotics, optimization, expert systems and financial applications (Hecht-Nielsen (1988b), Widrow *et al.* (1994)).

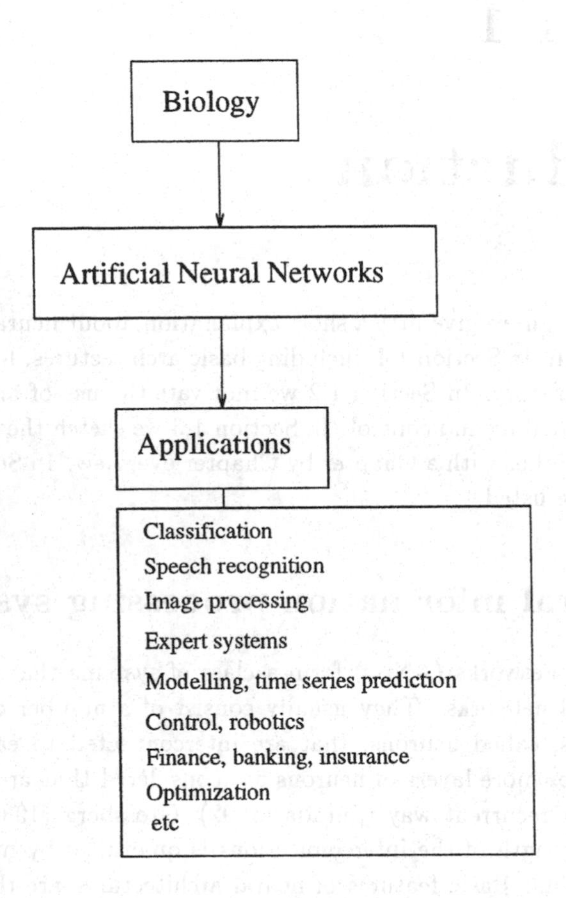


Figure 1.1: *The main source of inspiration for artificial neural networks is formed by the biological ones. In many domains neural nets have already proved to be very useful. In this book we focus on the theory, algorithms and applications for modelling and control of complex nonlinear systems.*

In trying to understand the emergence of the new discipline of neural networks, it is useful to look at some historical milestones:

- 1942 [McCulloch and Pitts]: simple static nonlinear model for a neuron, but no realizations.
- 1949 [Hebb]: first learning rule. One can memorize an object by adapting the weights.
- 1958 [Rosenblatt]: book on perceptrons, a machine that is capable of learning how to classify information by adapting the weights.
- 1960-1962 [Widrow & Hoff]: adalines and LMS rule
- 1969 [Minsky & Papert]: show theoretical limits of perceptrons as general computers.
- 23 years of hibernation but some 'stubborn' individuals (such as Grossberg, Amari, Fukushima, Kohonen, Taylor,...) continue doing research in neural nets.
- 1982 [Hopfield]: shows by means of energy functions that neural networks are capable of solving a large number of problems of the traveling salesman type. Re-emergence of the field.
- 1982 [Kohonen]: describes the self-organizing map.
- 1986 [Rumelhart]: rediscovers backpropagation.
- 1987 [Minsky]: knowledge systems evolve in blocks of specialised agents rather than as a homogeneous network (criticism in his book 'The Society of Mind').
- 1988 [Chua & Yang]: Cellular neural networks. Implementable networks, by taking neurons that are connected to their nearest neighbours only.
- Present: Blossoming international research community active in neural nets, as witnessed by the existence of several international neural net societies and many international conferences. Progress is continuously being made on the theory and practical applications.

One might state that the evolution of the field of neural networks is characterized by a number of ups and downs. The almost 25 years period of hibernation is due to the fact that networks without hidden layers were considered (perceptrons). Such models were unable e.g. to learn the well-known XOR problem, which is not a difficult problem in itself but cannot be represented by a single layer. A major breakthrough came by considering multilayer perceptrons, together with the backpropagation algorithm as its first learning paradigm. The fact that multilayer perceptrons are universal approximators, caused an explosion in artificial neural networks research and applications, which has now grown into a maturity.

| Year | Network | Inventor/Discoverer |
|---|---|---|
| 1942 | McCulloch-Pitts neuron | McCulloch, Pitts |
| 1957 | Perceptron | Rosenblatt |
| 1960 | Madaline | Widrow |
| 1969 | Cerebellatron | Albus |
| 1974 | Backpropagation network | Werbos, Parker, Rumelhart |
| 1977 | Brain state in a box | Anderson |
| 1978 | Neocognitron | Fukushima |
| 1978 | Adaptive Resonance Theory | Carpenter, Grossberg |
| 1980 | Self-organizing map | Kohonen |
| 1982 | Hopfield net | Hopfield |
| 1985 | Bidirectional assoc. mem. | Kosko |
| 1985 | Boltzmann machine | Hinton, Sejnowsky, Szu |
| 1986 | Counterpropagation | Hecht-Nielsen |
| 1988 | Cellular neural network | Chua, Yang |

Table 1.1: ***The best known artificial neural network architectures together with their year of introduction and their inventor/discoverer. See Hecht-Nielsen (1988) for part of this Table.***

Besides the many architectures (see Table 1.1) there are also differences in the types of learning mode. In supervised learning a training set of input/output data is given. In that case for each of these input patterns the network knows the desired output pattern. The network learns then to minimize the error between the actual output patterns and the desired ones, which can be done either off-line or on-line. In unsupervised learning desired output patterns are not available. The weights are adapted, using the input patterns only. The difference between supervised and unsupervised learning is visualized in Figure 1.2. Reinforcement learning is somewhere in between supervised and unsupervised learning. The network learns a task through a number of trials. A critic tells the network whether it is acting well or not and the network is rewarded or punished depending on the outcome of the trials.

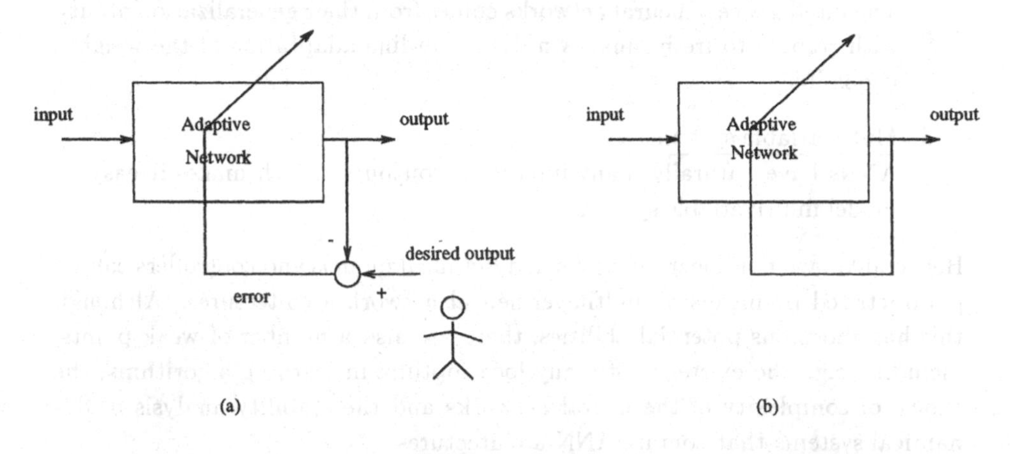


Figure 1.2: *Block diagram that illustrates the basic difference between: (a) supervised learning and (b) unsupervised learning. In supervised learning there is a teacher, somebody who tells the neural net how certain given inputs have to match certain desired outputs. This information will influence the way in which the synaptic (interconnection) weights are adapted. In unsupervised learning there is no external teacher.*

## 1.2 ANNs for modelling and control

The following features of artificial neural networks makes them particularly attractive and promising for application to modelling and control of nonlinear systems (see also Hunt *et al.* (1992)):

1. ANNs are universal approximators:
   It is proven that any continuous nonlinear function can be approximated arbitrarily well over a compact interval by a multilayer neural network that consists of one or more hidden layers.

2. Parallel distributed processing:
   The network has a highly parallel structure and consists of many processing elements with a very simple structure, which is interesting from the viewpoint of implementation.

3. Hardware implementation:
   Dedicated hardware is possible, resulting in additional speed.

4. Learning and adaptation:
   The intelligence of neural networks comes from their generalization ability with respect to fresh, unknown data. On-line adaptation of the weights is possible.

5. Multivariable systems:
   ANNs have naturally many inputs and outputs, which makes it easy to model multivariable systems.

Hence unknown nonlinear functions in dynamical models and controllers can be parametrized by means of multilayer neural network architectures. Although this has enormous potential abilities, there are also a number of weak points, including e.g. the existence of many local optima in learning algorithms, the choice of complexity of the neural networks and the stability analysis of dynamical systems that contain ANN architectures.

Moreover understanding the use of neural networks with respect to many types of existing model structures and control strategies is not a trivial task. Figure 1.3 gives some idea about the many theories and methods that are involved with this use of neural nets in identification and control. A new terminology emerged in the theory of neural networks, such as e.g. feedforward networks, recurrent networks, supervised learning, unsupervised learning, learning set, test set, generalization etc. One has the following links between the field of neural networks and control theory:

| **Neural Networks** | **Control Theory** |
| --- | --- |
| Feedforward net | Static nonlinear model |
| Recurrent net | Dynamic nonlinear model |
| Learning | Optimization |
| Training set | I/O data for identification |
| Test set | Fresh data |
| Generalization | Cross validation |

In this book ANNs for modelling and control will be studied from a system theoretical viewpoint.

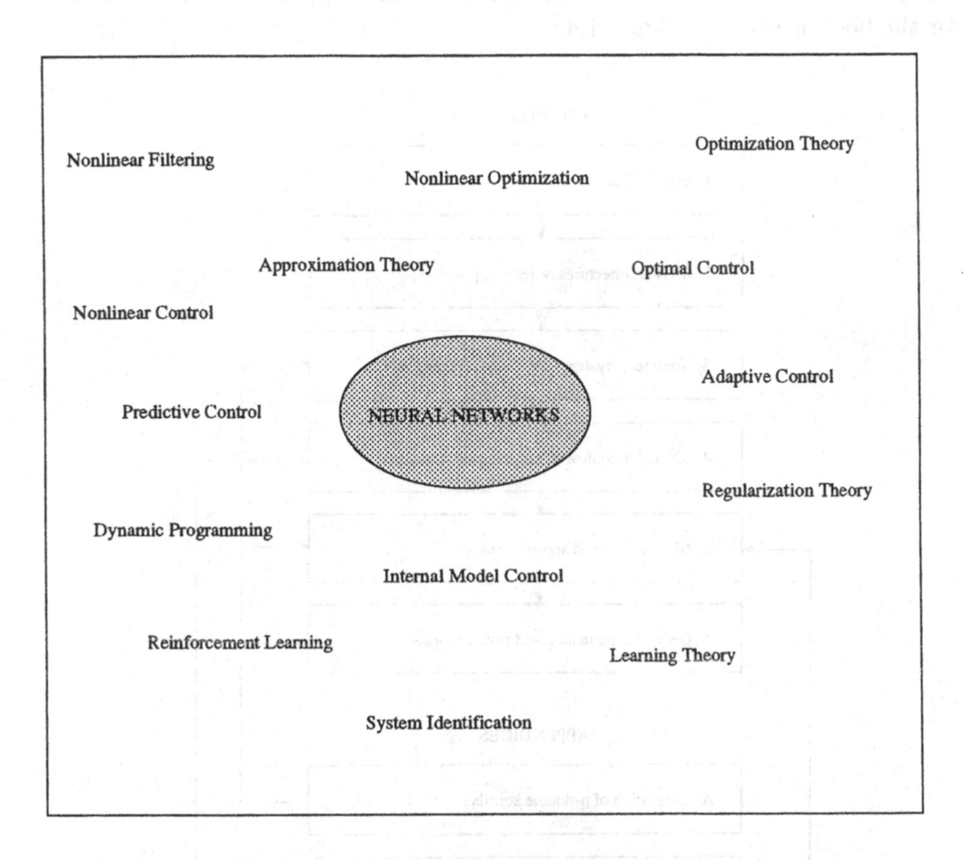


Figure 1.3: *The many mathematical engineering disciplines that are related to neural networks in modelling and control. Sometimes, it is far too easily assumed that neural nets could offer magic solutions to difficult problems. Neural network architectures are powerful, but non-trivial to analyse because nonlinearities, that are classically called 'hard', are involved.*

## 1.3 Chapter by Chapter overview

We give here a short overview of the Chapters and Appendices. A flow chart for the book is given in Figure 1.4.

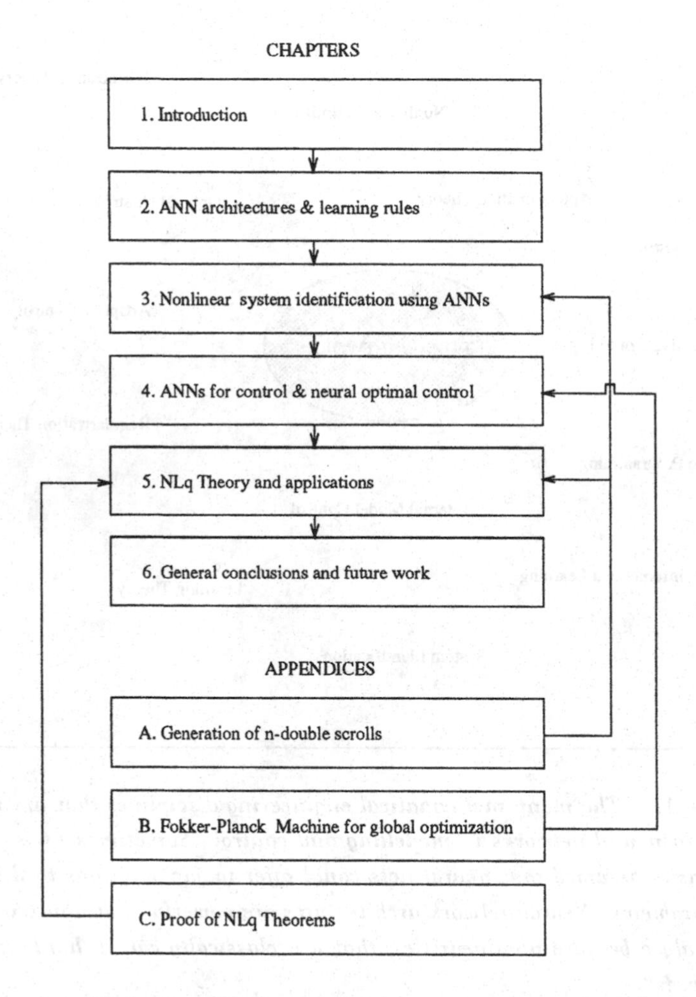


Figure 1.4: *Flow chart for this book. After giving a review of basic ANN architectures and classical learning rules in Chapter 2, main contributions concerning nonlinear system identification, neural optimal control and stability analysis of neural control systems are stated in Chapter 3 to 5. Links, similarities and differences with existing work are discussed.*

## Chapter 2. ANNs: architectures and learning rules

In this Chapter we discuss multilayer perceptrons and radial basis function networks. After giving a network description, universal approximation theorems are stated for the networks. The intriguing history of multilayer perceptrons is developed, starting from Kolmogorov's theorem. Theorems by several authors such as Cybenko, Funahashi and Hornik state that any continuous nonlinear function can be approximated arbitrarily well by a multilayer neural network with one or more hidden layers. Moreover Barron proved that multilayer perceptrons may avoid the curse of dimensionality, which is a considerable advantage over polynomial expansions. For radial basis function networks it is shown according to Poggio & Girosi that they have a best representation property. The networks are closely related to the theory of splines.

Then classical paradigms of learning are reviewed. For multilayer feedforward networks this is the backpropagation algorithm. For recurrent neural networks one has to apply Narendra's dynamic backpropagation or backpropagation through time of Werbos. In gradient based learning algorithms for recurrent nets, the gradient is generated by a system which is in itself dynamical, like the recurrent net. In dynamic backpropagation this is called a sensitivity model. Off-line learning is interpreted as solving a non-convex optimization problem.

Finally the use of linear regression for feedforward RBF networks is motivated, based upon Poggio's theory. A suboptimal procedure is to place first the centers using unsupervised learning and to apply linear regression then in order to find the output weights. The emphasis in this book is however on the use of multilayer perceptrons.

## Chapter 3. Nonlinear system identification using ANNs

First a short overview of linear dynamical model structures is given and their extension towards nonlinear dynamical models is explained, with input/output models such as NARX and NARMAX as well as nonlinear state space models. Then parametrizations of the model structures by multilayer perceptrons are made. For I/O models the difference between series-parallel and parallel mode is explained according to Narendra. Series-parallel mode leads to feedforward networks and parallel mode to recurrent networks. Neural state space models are introduced that take into account process noise and measurement noise. The predictor is in innovations form. Instead of taking into account the dependency of the Kalman gain through a Riccati equation as would follow from extended Kalman filtering, a direct parametrization of the steady state

Kalman gain is made. The filter has then a similar format as the Kalman filter for linear systems. Identifiability of neural state space models is investigated, based upon the work of Sussmann and Albertini & Sontag. This shows that neural state space models are unique up to a similarity transformation and sign reversals.

An overview of learning algorithms for feedforward and recurrent network related models is presented. Both off-line (batch) and on-line (recursive) algorithms are explained. For off-line identification the emphasis is on prediction error algorithms. Both classical paradigms of learning such as static and dynamic backpropagation as well as advanced algorithms are discussed. Advanced algorithms, including quasi-Newton, Levenberg-Marquardt and conjugate gradient methods, follow from nonlinear optimization theory. For on-line identification recursive prediction error algorithms and extended Kalman filtering are adopted. Special attention is given on the calculation of the gradient of the cost function for neural state space models. The sensitivity model is derived, according to Narendra's dynamic backpropagation method. Aspects of model validation, pruning and regularization are given.

Neural network models are interpreted as uncertain linear systems. This brings a better insight into the black box structure of such models. Some consequences are the possibility for more specific parametrizations, estimating hardness of distortion of the underlying nonlinearities of the identified model and adopting results from linear identification as starting points for neural network models. Moreover, for neural state space models a Linear Fractional Transformation (LFT) representation is derived. As a result identified neural state space models fit immediately into the framework of modern robust control theory.

Finally, examples on nonlinear system identification are given. Some convincing examples from literature for I/O models include e.g. the Santa Fe time series prediction competition, the modelling of the space shuttle main engine and electric load forecasting. Both simulated and real life examples are given for neural state space models: a simulated nonlinear system with hysteresis, corrupted by process noise and measurement noise, identification of a glass furnace and identification of Chua's double scroll (chaos) by means of simple recurrent nets.

## Chapter 4. Neural networks for control

An introduction on neural control is given first. Basic principles of direct and indirect neural adaptive control, reinforcement learning, neural optimal control, internal model control and model predictive control using neural networks are discussed. Hence neural control strategies can be either model-based or not.

The emphasis in this Chapter is on neural optimal control. Starting from the $N$-stage optimal control problem in nonlinear optimal control, it is explained how optimal control problems can be solved efficiently by parametrizing unknown nonlinear functions by means of neural network architectures. Both the stabilization problem and tracking problem with static state feedback are explained according to Saerens and Parisini & Zoppoli respectively. Then static and dynamic output feedback laws are considered and suboptimal solutions by means of parametric optimization problems are formulated. Like in identification problems with recurrent nets, Narendra's dynamic backpropagation plays again an important role in calculating the derivative of the cost function. Furthermore it is shown how results from linear control theory can be incorporated into the neural control design. Because neural controllers have additional degrees of freedom with respect to linear controllers, due to the hidden layer, they can do more than local stabilization at a target point. It is illustrated how transitions between equilibrium points with local stability at the target points can be realized using neural control, despite the finite time horizon that is considered in the optimization problems. Examples are the solution to the swinging up problem of an inverted and double inverted pendulum system, both by feedforward and recurrent neural networks. Local optimization methods like Sequential Quadratic Programming as well as global optimization by means of genetic algorithms have been applied. Excellent results on the inverted pendulum problem have been obtained by means of a Fokker-Planck learning machine, further explained in Appendix B. Furthermore it turns out that small scale neural networks are able to solve complicated control tasks.

## Chapter 5. NL$_q$ theory

While in the previous Chapter neural control has been tackled from the viewpoint of optimal control theory, in this Chapter we take the viewpoint of modern (robust) control theory. A model based framework with nonlinear state space models and controllers, parametrized by multilayer perceptrons and simply called neural state space models and controllers, is developed. Major contributions in this Chapter are sufficient stability criteria for global asymptotic stability and input/output stability with finite $L_2$-gain for closed loop systems. Therefore so-called NL$_q$ systems were introduced. NL$_q$s represent a large class of nonlinear dynamical systems in state space form and contain a number of $q$ layers of an alternating sequence of linear and static nonlinear operators that satisfy a sector condition. All system descriptions are transformed into this $NL_q$ form and stability criteria are presented for $NL_q$s. Three types of stability criteria are derived: Theorems on diagonal scaling, Theorems on diagonal dominance and Theorems expressed in terms of condition number factors of certain matrices. For a given controller (or fixed matrices in the $NL_q$) checking these criteria corresponds to solving a convex optimization problem. Indeed the criteria can be cast as LMIs (Linear Matrix Inequalities). The diagonal scaling criteria already suggest some connections with $\mu$ theory in robust control. The precise links with $\mu$ theory are investigated by considering perturbed $NL_q$s: it is shown then under what conditions Packard & Doyle's state space upper bound test for robust performance is a special case of $NL_q$ theorems for $q = 1$ (in fact the most conservative theorems on diagonal scaling).

After presenting the main Theorems, neural control design within $NL_q$ theory is explained. It leads to the formulation of a modified dynamic backpropagation algorithm in order to solve tracking problems. $NL_q$ theorems are used in order to assess closed loop stability. Objective functions with singular value constraints are to be considered which results in non-convex nondifferentiable optimization. Even when it's impossible to find a feasible point to the $NL_q$ theorems, their principles are still valuable in order to enforce complex nonlinear behaviour towards global asymptotic stability. This is done by imposing local stability and making certain matrices as diagonally dominant as possible or by minimizing a certain condition number factor. The latter has been illustrated e.g. on mastering chaos in a neural state space model emulator for Chua's double scroll attractor. Furthermore, enforcing systems that have multiple equilibria, are periodic or even chaotic towards global asymptotic stability is possible. A final example is controlling nonlinear distortion in electrodynamic loudspeakers, using the model based approach within $NL_q$ theory.

Finally it turns out that $NL_q$s have a unifying nature, in the sense that many problems arising in neural networks, systems and control can be interpreted as special cases. Examples of recurrent neural networks that are represented as special cases of $NL_q$s are e.g. multilayer Hopfield networks, Generalized CNNs and locally recurrent globally feedforward networks. The latter is in itself a generalization of many type of existing recurrent network architectures. An overview of examples on $NL_q$s and related theories is given in Table 1.2 and Figure 1.5.

| $NL_q$ system | $q$ value | Application |
|---|---|---|
| Neural state space control systems | $q \geq 1$ | modelling and control |
| Generalized CNNs | $q \geq 1$ | signal processing |
| LFTs with real diagonal $\Delta$ block | $q = 1$ | control |
| Lur'e problem | $q = 1$ | control |
| Linear control scheme with saturation input | $q = 1$ | control |
| Digital filters with overflow characteristic | $q = 1$ | signal processing |
| Hopfield network, CNN | $q = 1$ | image processing |
| LRGF networks | $q = 1$ | modelling |

Table 1.2: ***Special cases of $NL_q$s arising in neural networks, signal processing, modelling and control.***

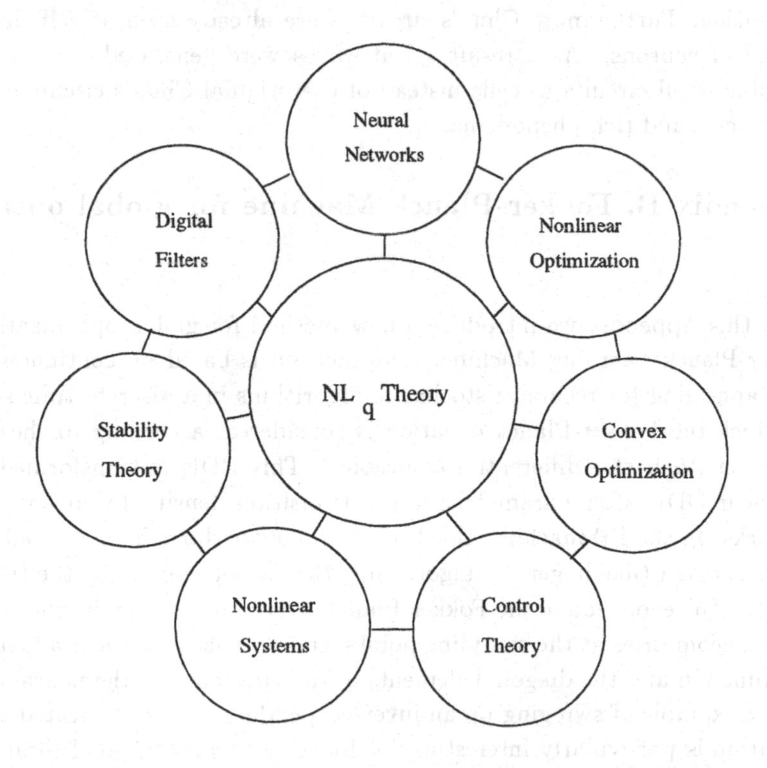


Figure 1.5: ***$NL_q$ theory and its relation to other mathematical engineering disciplines.***

## Appendix A. Generation of $n$-double scrolls

In this Appendix we discuss a generalization to Chua's circuit, which is well-known to be a paradigm for chaos in electrical circuits. This is done by introducing additional break points in the characteristic of the nonlinear resistor, leading to $n$-double scrolls. The new circuit is a generalization in the sense that the 1-double scroll corresponds to the classical double scroll. Depending on a bifurcation parameter of the circuit one goes from multiple sink to multiple scroll portraits, like one bifurcates from a sink portrait to a double scroll in the original circuit. Identification and control of a neural state space model emulator for the double scroll was done in Chapter 3 and 5 respectively. From extrapolation of current applications of Chua's circuit, one might expect possible applications and new results for the $n$-double scroll circuit e.g. in secure communication, music and modelling of chaotic dynamics in neural pattern recognition. Furthermore Chua's circuits were already used as cells in CNNs instead of neurons. As a result spiral waves were generated. Hence taking $n$-double scroll circuits as cells instead of the original Chua's circuit may also lead to new and rich phenomena.

## Appendix B. Fokker-Planck Machine for global optimization

In this Appendix we introduce a new method for global optimization, the Fokker-Planck Learning Machine. The method is based on continuous simulated annealing (or recursive stochastic algorithms in a discrete time setting), for which the Fokker-Planck equation is considered, according to the general theory of stochastic differential equations. This PDE is transformed into a nonlinear ODE after parametrizing the transition density by means of RBF networks. In the FP machine, this leads to an algorithm where a population of points is taken (like in genetic algorithms) that is determined by the transition density. The evolution of the Fokker-Planck equation is driven by the ensemble of local geometries at the sampling points, characterized by the gradient of the cost function and the diagonal elements of the Hessian. On the neural optimal control example of swinging up an inverted pendulum it is illustrated that the algorithm is particularly interesting for locating regions of good local optima and that it is especially successful in combination with a fast local optimization scheme. Furthermore the example also shows that the FP machine is more effective than multistart local optimization.

### Appendix C. Proof of $NL_q$ Theorems

This Appendix contains the proofs of $NL_q$ Theorems. Essential aspects in the proofs are the use of radially unbounded Lyapunov functions (in order to prove global asymptotic stability), the fact that $NL_q$ systems are in state space form, the use of vector norms as Lyapunov functions and properties of induced norms.

## 1.4 Contributions

In this Section we give a list of our new contributions to the field of neural networks, in order to enable a better discrimination of these new results within the broader context of this book:

- *Neural state space models:*
  The neural state space model is a new type of predictor, which results from parametrizing nonlinear state space models by means of multilayer perceptrons. The model contains a steady state Kalman gain. The links with extended Kalman filtering are discussed. Aspects of identifiability are considered. Learning algorithms are discussed based on Narendra's dynamic backpropagation, because neural state space models are recurrent neural networks. Expressions for the gradient are generated by sensitivity models. Neural state space models were introduced in Suykens *et al.* (1994b).

- *Neural network models as uncertain linear systems:*
  Neural network models are interpreted as linear models with state and input dependent system matrices. Moreover for neural state space models an LFT representation is derived, such that identified models fit immediately in the framework of robust control theory. Some other consequences are a better insight into the behaviour and working principles of black box neural state space models, the use of results from linear identification as a priori information for initializing neural network models and the estimation of parametric uncertainties. References for neural network models as uncertain linear systems are Suykens & De Moor (1993b,e), Suykens *et al.* (1994b).

- *Neural optimal control constrained by results from linear control:*
  It is shown that neural controllers are able to do more than local stabilization. They have additional degrees of freedom with respect to linear controllers. Results from linear control theory can be used as constraints

on the choice of the interconnection weights. Moreover it is possible to formulate an unconstrained problem by eliminating the constraint. Nonlocal control such as realizing transitions between equilibrium points is possible. Neural optimal control problems are formulated as parametric optimization problems, which were solved by local optimization methods as well as global optimization methods including genetic algorithms and the Fokker-Planck machine. Although a finite time horizon is considered in the cost function, it is possible to achieve local stability at the target point, due to the linear controller design constraint. Both feedforward and recurrent neural nets are used for static and dynamic output feedback respectively. It is shown by examples that small scale neural nets are able to solve complicated control problems. References are Suykens *et al.* (1993c)(1994a).

- *$NL_q$ theory:*
  A model-based theory for top-down neural control design is introduced. The framework consists of neural state space models and controllers, considered in standard plant form according to modern control theory. In order to analyse these systems, the $\mathrm{NL}_q$ system form is introduced. Sufficient conditions for global asymptotic stability, input/output stability and robust performance are derived. Connections between $\mathrm{NL}_q$ theory and robust control theory ($H_\infty$ and $\mu$ theory) are revealed. In fact the latter correspond to the special case $q = 1$. Stability criteria are formulated as LMIs. A modified dynamic backpropagation is presented in order to obtain optimal tracking problems, under the constraint of closed loop stability. Recurrent neural network architectures such as the (multilayer) Hopfield networks, (generalized) CNNs and LRGFs are represented as $\mathrm{NL}_q$s. References for $\mathrm{NL}_q$ theory are Suykens *et al.* (1994c,d,f,g,h), Suykens & Vandewalle (1994e)(1995a).

- *Generation of n-double scrolls:*
  By modifying the characteristic of the nonlinear resistor in Chua's circuit, $n$-double scrolls are generated. Bifurcations from multiple sink to multiple scroll portraits are obtained. References are Suykens & Vandewalle (1991)(1993a,d).

- *Fokker-Planck Learning Machine:*
  The FP machine is a new algorithm for global optimization which follows from considering the Fokker-Planck equation for continuous simulated annealing or recursive stochastic algorithms. Parametrization of the transition density by means of RBF networks transforms the PDE

into a nonlinear ODE. The algorithm works with a population of points like genetic algorithms. References are Suykens & Vandewalle (1994e,i).

These theoretical results are illustrated by a number of examples and new applications. Among the examples are:

- *Identification of a glass furnace:*
  This is a real life example on nonlinear system identification using neural state space models. Results from linear identification served as a starting point. Improved fitting and generalization have been obtained using neural networks.

- *Identifying n-double scrolls:*
  A simple recurrent neural network emulator for Chua's double scroll is identified using neural state space models. A special training strategy is required. A 2-double scroll is identified using I/O models in series-parallel mode.

- *Swinging up an inverted and double inverted pendulum:*
  The use of results from linear control theory for neural optimal control design by means of feedforward and recurrent neural nets is illustrated on the example of swinging up an inverted and double inverted pendulum.

- *Towards global stabilization of any type of nonlinear behaviour:*
  It is shown how nonlinear behaviour with multiple equilibrium points, periodic behaviour or chaos, revealed by neural state space model, can be stabilized towards global asymptotic stability by means of $NL_q$ theory.

- *Mastering chaos from a double scroll emulator:*
  The chaotic behaviour of the double scroll emulator is stabilized and controlled within $NL_q$ theory.

- *Controlling nonlinear distortion in electrodynamic loudspeakers:*
  In electrodynamic loudspeakers unwanted harmonics and intermodulation products are produced at low frequency, due to several nonlinearities. A model-based neural network approach within $NL_q$ theory is proposed in order to linearize the system.

In this book an attempt is made to situate these new results within the general context of neural networks for modelling and control.

# Chapter 2

# Artificial neural networks: architectures and learning rules

In this Chapter we discuss two basic types of artificial neural network architectures that are used in the sequel for modelling and control purposes: the multilayer perceptron and the radial basis function network. This Chapter is organized as follows. In Section 2.1 we give a description of the architectures. In Section 2.2 we present an overview of universal approximation theorems, together with a brief historical context. In Section 2.3 classical learning paradigms for feedforward and recurrent neural networks and RBF networks are reviewed.

## 2.1 Basic neural network architectures

The best known neural network architecture is the multilayer feedforward neural network (multilayer perceptron). It is a static network that consists of a number of layers: input layer, output layer and one or more hidden layers connected in a feedforward way (see e.g. Zurada (1992)). Each layer consists of a number of McCulloch-Pitts neurons. One single neuron makes the simple operation of a weighted sum of the incoming signals and a bias term (or threshold), fed through an activation function $\sigma(.)$ and resulting in the output value of the neuron. A network with one hidden layer is described in matrix-vector notation as

$$y = W\,\sigma(Vx + \beta), \tag{2.1}$$

or in elementwise notation:

$$y_i = \sum_{r=1}^{n_h} w_{ir}\,\sigma(\sum_{j=1}^{m} v_{rj} x_j + \beta_r), \qquad i = 1, ..., l. \tag{2.2}$$

Here $x \in \mathbb{R}^m$ is the input and $y \in \mathbb{R}^l$ the output of the network and the nonlinear operation $\sigma(.)$ is taken elementwise. The interconnection matrices are $W \in \mathbb{R}^{l \times n_h}$ for the output layer, $V \in \mathbb{R}^{n_h \times m}$ for the hidden layer, $\beta \in \mathbb{R}^{n_h}$ is the bias vector (thresholds of hidden neurons) with $n_h$ the number of hidden neurons.

For a network with two hidden layers (see Figure 2.1) one has

$$y = W\,\sigma(V_2\,\sigma(V_1 x + \beta_1) + \beta_2), \tag{2.3}$$

or

$$y_i = \sum_{r=1}^{n_{h_2}} w_{ir}\,\sigma(\sum_{s=1}^{n_{h_1}} v_{rs}^{(2)}\,\sigma(\sum_{j=1}^{m} v_{sj}^{(1)} x_j + \beta_s^{(1)}) + \beta_r^{(2)}), \qquad i = 1, ..., l. \tag{2.4}$$

The interconnection matrices are $W \in \mathbb{R}^{l \times n_{h_2}}$ for the output layer, $V_2 \in \mathbb{R}^{n_{h_2} \times n_{h_1}}$ for the second hidden layer and $V_1 \in \mathbb{R}^{n_{h_1} \times m}$ for the first hidden layer. The bias vectors are $\beta_2 \in \mathbb{R}^{n_{h_2}}$, $\beta_1 \in \mathbb{R}^{n_{h_1}}$ for the second and first hidden layer respectively. In order to describe a network with $L$ layers ($L - 1$ hidden layers, because the input layer is a 'dummy' layer), the following notation will be used in the sequel

$$x_i^l = \sigma(\xi_i^l), \qquad \xi_i^l = \sum_{j=1}^{N_l} w_{ij}^l\, x_j^{l-1} \tag{2.5}$$

where $l = 1, ..., L$ is the layer index, $N_l$ denotes the number of neurons in layer $l$ and $x_i^l$ is the output of the neurons at layer $l$. The thresholds are considered here to be part of the interconnection matrix, by defining additional constant inputs.

The choice of the activation function $\sigma(.)$ may depend on the application area. Typical activation functions are shown in Figure 2.2. For applications in modelling and control the hyperbolic tangent function $\tanh(x) = (1-\exp(-2x))/(1+\exp(-2x))$ is normally used. In case of a 'tanh' the derivative of the activation function is $\sigma' = 1 - \sigma^2$. The neurons of the input layer have a linear activation function.

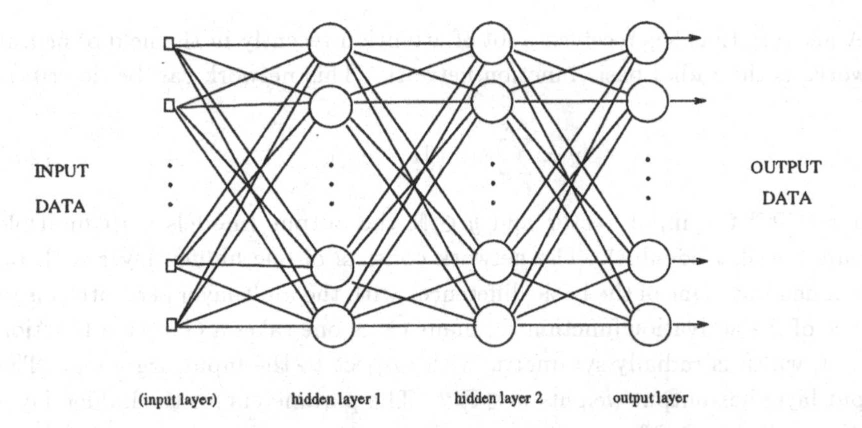


Figure 2.1: *This figure shows a multilayer perceptron, which is a static nonlinear network that consists of a dummy input layer, an output layer and two hidden layers. A layer consists of a number of McCulloch-Pitts neurons that perform the operation of a weighted sum of incoming signals, feeded through a saturation-like nonlinearity. One hidden layer is sufficient in order to be a universal approximator for any continuous nonlinear function.*

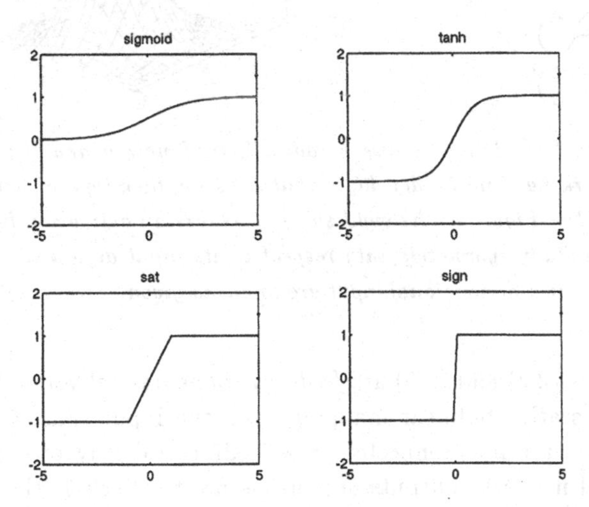


Figure 2.2: *Some possible activation functions for the neurons in the multilayer perceptron. In this book we take the hyperbolic tangent function* tanh(.). *Note that this is a static nonlinearity that belongs to the sector* $[0, 1]$.

A network that has received a lot of attention recently in the field of neural networks is the radial basis function network. This network can be described as

$$y = \sum_{i=1}^{n_h} w_i \, g(\|x - c_i\|), \tag{2.6}$$

with $x \in \mathbb{R}^m$ the input vector and $y \in \mathbb{R}$ the output (models with multiple outputs are also possible). The network consists of one hidden layer with $n_h$ hidden neurons. One of the basic differences with the multilayer perceptron is in the use of the activation function: in many cases one takes a Gaussian function for $g(.)$, which is radially symmetric with respect to the input argument. The output layer has output weights $w \in \mathbb{R}^{n_h}$. The parameters for the hidden layer are the centers $c_i \in \mathbb{R}^m$.

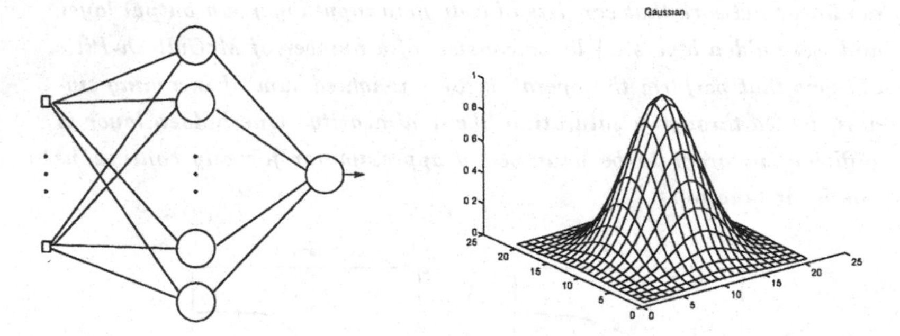


Figure 2.3: *This Figure shows a radial basis function neural network. The network has one hidden layer, but uses another type of activation function. The Figure at the right shows a Gaussian activation function, which is radially symmetric with respect to its input argument. An example for a two-dimensional input argument is given.*

The networks (2.3) and (2.6) are *feedforward* neural networks, which means that there is a static nonlinear mapping from the input space to the output space: the layers are interconnected in a feedforward way to each other. In *recurrent* neural networks outputs of neurons are fed back to the network, resulting in a dynamical system. A simple example of a recurrent neural network is the Hopfield network, in discrete time looking as

$$x_{k+1} = \tanh(W x_k) \tag{2.7}$$

with $x_k \in \mathbb{R}^n$ the state vector and $W \in \mathbb{R}^{n \times n}$ the synaptic matrix. The network consists of one layer of neurons. Throughout this work we take synchronous updating of the neurons.

## 2.2 Universal approximation theorems

In this Section we will discuss approximation theorems for multilayer feedforward neural networks and radial basis function networks. A quick glance is given on the intriguing history and some recent Theorems are presented.

### 2.2.1 Multilayer perceptrons

The history of universal approximation by neural networks starts in fact in the year 1900, when Hilbert formulated a list of 23 challenging problems for the century to come. The famous 13-th problem is the following conjecture (see Funahashi (1989)).

**Conjecture [Hilbert, 1900].** There are analytic functions of three variables which cannot be represented as a finite superposition of continuous functions of only two variables.

This conjecture was refuted by Kolmogorov and Arnold in 1957. They proved the following Theorem:

**Theorem 2.1 [Kolmogorov, 1957].** Any continuous function $f(x_1, ..., x_n)$ of several variables defined on the cube $[0,1]^n$ $(n \geq 2)$ can be represented in the form

$$f(x) = \sum_{j=1}^{2n+1} \chi_j(\sum_{i=1}^{n} \psi_{ij}(x_i))$$

where $\chi_j, \psi_{ij}$ are continuous functions of one variable and $\psi_{ij}$ are monotone functions which are not dependent on $f$.

□

This Theorem was refined by Sprecher as follows:

**Theorem 2.2 [Sprecher, 1965].** For each integer $n \geq 2$, there exists a real, monotone increasing function $\psi(x)$, $\psi([0,1]) = [0,1]$, depending on $n$ and having the property: for each preassigned number $\delta > 0$ there is a rational number $\epsilon$, $0 < \epsilon < \delta$, such that every real continuous function of $n$ variables $f(x)$, defined on $[0,1]^n$, can be represented as

$$f(x) = \sum_{j=1}^{2n+1} \chi[\sum_{i=1}^{n} \lambda^i \psi(x_i + \epsilon(j-1)) + j - 1]$$

where the function $\chi$ is real and continuous and $\lambda$ is a constant independent of $f$.

□

A link between the Sprecher Theorem and neural networks was first revealed by Hecht-Nielsen in 1987. He pointed out that the Sprecher Theorem means that any continuous mapping $f : x \in [0,1]^n \rightarrow (f_1(x), ..., f_m(x)) \in \mathbb{R}^m$ is represented by a form of a two hidden layer neural network with hidden units whose output functions are $\psi$, $\chi_i (i = 1, .., m)$, where $\psi$ is used for the first hidden layer, $\chi_i$ are used for the second hidden layer and is given by the Sprecher Theorem for $f_i(x)$ (see Funahashi (1989), Hecht-Nielsen (1987)). However the fact that the functions $\psi_{ij}$ are highly nonsmooth and the functions $\chi_i$ depend on the specific function $f$ and are not representable in a parametrized form, was part of the criticism made by Girosi & Poggio, who claimed that Kolmogorov's representation theorem is irrelevant for neural networks (Girosi & Poggio (1989)). In 1991 this was refuted by Kůrková, who proved that Kolmogorov's Theorem is indeed relevant: by using staircase-like functions of the form $\sum_i a_i \sigma(b_i x + c_i)$ , where $\sigma(.)$ is a sigmoidal function, for $\chi_i$ and $\psi_{ij}$ in the two hidden layer network it is indeed possible to approximate any continuous function arbitrarily well (Kůrková (1991)(1992)).

On the other hand in 1989 it was also shown independently by Hornik *et al.* (1989), Funahashi (1989) and Cybenko (1989) that a multilayer feedforward neural network with one or more hidden layers is sufficient in order to approximate any continuous nonlinear function arbitrarily well on a compact interval, provided sufficient hidden neurons are available. In contrast to Kůrková, they make use of advanced theorems from functional analysis. We will focus here on the results presented in Hornik (1989). In order to understand the following Theorems some preliminary definitions have to be introduced. Let $C^r$ denote the set of continuous functions $\mathbb{R}^r \rightarrow \mathbb{R}$, $\rho$ a metric to measure the distance between $f, g \in C^r$, $A^r$ the set of all affine functions from $\mathbb{R}^r$ to $\mathbb{R}$: $\{A(x) : \mathbb{R}^r \rightarrow \mathbb{R} : A(x) = w^T x + b;\ x \in \mathbb{R}^r\}$. So called $\sum$ networks and $\sum\prod$ networks are then defined as

$$\begin{aligned} \textstyle\sum^r(G) &= \{f : \mathbb{R}^r \rightarrow \mathbb{R} : f(x) = \textstyle\sum_{j=1}^q \beta_j\, G(A_j(x)); x \in \mathbb{R}^r, \beta_j \in \mathbb{R}, A_j \in A^r\} \\ \textstyle\sum\prod^r(G) &= \{f : \mathbb{R}^r \rightarrow \mathbb{R} : f(x) = \textstyle\sum_{j=1}^q \beta_j \prod_{k=1}^{l_j} G(A_{jk}(x)); \\ & \qquad x \in \mathbb{R}^r, \beta_j \in \mathbb{R}, A_{jk} \in A^r\} \end{aligned}$$

where $G : \mathbb{R} \rightarrow \mathbb{R}$ is a Borel measurable function. A function $\psi : \mathbb{R} \rightarrow [0,1]$ is a squashing function if $\psi$ is nondecreasing, $\lim_{\lambda \rightarrow -\infty} \psi(\lambda) = 0$ and $\lim_{\lambda \rightarrow \infty} \psi(\lambda) = 1$ and has a countable number of discontinuities. Remark that a $\sum^r(G)$ network reduces to a standard classical multilayer feedforward neural network with one hidden layer and activation function $\psi$ if $G = \psi$ ($\sum^r(\psi)$ network). Then we need the following definitions:

**Definition 2.1.** A subset $S$ of a metric space $(X, \rho)$ is $\rho$-dense in a subset $T$ if $\forall \epsilon > 0, \forall t \in T, \exists s \in S$ *s.t.* $\rho(s,t) < \epsilon$.

**Definition 2.2.** $S \subset C^r$ is uniformly dense on compacta in $C^r$ if for all compact subsets $K \subset \mathbb{R}^r$ the subset $S$ is $\rho_K$-dense in $C^r$ with $\rho_K(f,g) = \sup_{x \in K} |f(x) - g(x)|$. A sequence $\{f_n\}$ converges to $f$ uniformly on compacta if for all $K \subset \mathbb{R}^r$: $\rho_K(f_n, f) \to 0$ as $n \to \infty$.

In our case $T$ corresponds to $C^r$ and $S$ to $\sum^r(G)$ or $\sum\prod^r(G)$. The following theorem then holds:

**Theorem 2.3 [Hornik, 1989].** Let $G$ be any continuous nonconstant function from $\mathbb{R}$ to $\mathbb{R}$. Then $\sum\prod^r(G)$ is uniformly dense on compacta in $C^r$.

□

Hence $\sum\prod(G)$ feedforward networks are capable of arbitrarily accurate approximation to any real-valued continuous function over a compact set and $G$ may be *any* continuous nonconstant function here.

Now let $\mu$ be a probability measure defined on $(\mathbb{R}^r, B^r)$ with $B^r \subset \mathbb{R}^r$ a Borel $\sigma$-field and $M^r$ the set of all Borel measurable functions from $\mathbb{R}^r$ to $\mathbb{R}$. Functions $f, g \in M^r$ are called $\mu$-equivalent if $\mu\{x \in \mathbb{R}^r : f(x) = g(x)\} = 1$. Then the metric $\rho_\mu : M^r \times M^r \to \mathbb{R}^+$ is defined by $\rho_\mu(f,g) = \inf\{\epsilon > 0 : \mu\{x : |f(x) - g(x)| > \epsilon\} < \epsilon\}$. Hence $f$ and $g$ are close in the metric $\rho_\mu$ if and only if there is only a small probability that they differ significantly and $f$ and $g$ are $\mu$-equivalent if $\rho_\mu(f,g) = 0$. The following Theorem then holds

**Theorem 2.4 [Hornik, 1989].** For every squashing function $\psi$ and every probability measure $\mu$ on $(\mathbb{R}^r, B^r)$, $\sum^r(\psi)$ is uniformly dense on compacta in $C^r$ and $\rho_\mu$-dense in $M^r$.

□

This means that regardless of the dimension $r$ of the input space and for any squashing function $\psi$, a feedforward neural network with one hidden layer can approximate *any* continuous function arbitrarily well in the $\rho_\mu$ metric. In the proof of Hornik's Theorems a central role is played by the Stone-Weierstrass theorem.

In addition to the previous Theorems, more refined Theorems were formulated by Hornik (1991). More recently Leshno *et al.* (1993) showed that a standard multilayer feedforward network with a locally bounded piecewise continuous activation function can approximate any continuous function to any degree of accuracy if and only if the network's activation function is not a polynomial.

The previous Theorems are existence theorems and are not constructive in the sense that no learning algorithms are presented and they give no or little idea about the number of hidden units to be used for the approximation. Furthermore no comparison is made between neural networks and other universal approximators such as polynomial expansions in terms of network complexity. The latter problem is addressed by Barron (1993). It has been shown that the parsimony of a neural network parametrization is surprisingly advantageous in high-dimensional settings. Feedforward networks with one hidden layer of sigmoidal activation function achieve an integrated squared error of order $O(1/n)$, independent of the dimension of the input space, where $n$ denotes the number of hidden neurons. The underlying function to be approximated is assumed to have a bound on the first moment of the magnitude distribution of the Fourier transform (smoothness property). On the other hand for a series expansion with $n$ terms, in which only the parameters of a linear combination are adjusted (such as traditional polynomial, spline and trigonometric expansions), the integrated square error cannot be made smaller than $O(1/n^{2/d})$ where $d$ is the dimension of the input, for functions satisfying the same smoothness assumption. In order to formulate Barron's theorem, let $f(x)$ denote the class of functions on $\mathbb{R}^d$ for which there is a Fourier representation of the form $f(x) = \int_{\mathbb{R}^d} \exp(i\omega x)\ \tilde{f}(\omega)\ d\omega$, for some complex-valued function $\tilde{f}(\omega)$ for which $\omega \tilde{f}(\omega)$ is integrable, and define $C_f = \int_{\mathbb{R}^d} |\omega|\ |\tilde{f}(\omega)|\ d\omega$. Let $f_n = \sum_{k=1}^{n} c_k\ \psi(a_k x + b_k) + c_0$ with $a_k \in \mathbb{R}^d$, $b_k, c_k \in \mathbb{R}$ denote a linear combination of sigmoidal functions with squashing function $\psi$. Furthermore let $\mu$ denote a probability measure on the ball $B_r = \{x : |x| \leq r\}$ with radius $r > 0$. Then the following holds

**Theorem 2.5 [Barron, 1993].** For every function $f$ with $C_f$ finite, and every $n \geq 1$, there exists a linear combination of sigmoidal functions $f_n(x)$, such that

$$\int_{B_r} (f(x) - f_n(x))^2\ \mu(dx) \leq \frac{k_f}{n}$$

where $k_f = (2rC_f)^2$.

□

Hence the effects of the curse of dimensionality are avoided in terms of the accuracy of approximation.

### 2.2.2 Radial basis function networks

Like for multilayer perceptrons universal approximation theorems are also available for radial basis function neural networks. Park & Sandberg (1991) showed the following result.

**Theorem 2.6 [Park & Sandberg, 1991].** Let $S_K$ be the family of RBF networks, consisting of the function $q : \mathbb{R}^m \rightarrow \mathbb{R}$, represented by

$$q(x) = \sum_{i=1}^{n_h} w_i K(\frac{x - c_i}{\sigma_i})$$

where $K$ is a radially symmetric kernel function of a unit in the one hidden layer, $c_i \in \mathbb{R}^m$ is the centroid and $\sigma_i$ the smoothing factor (or width) of the $i$th kernel node. Now let $K : \mathbb{R}^m \rightarrow \mathbb{R}$ be an integrable bounded function such that $K$ is continuous almost everywhere and $\int_{\mathbb{R}^m} K(x)dx \neq 0$ and let $\sigma_i = \sigma$ for all $n_h$ hidden units. Then the family $S_K$ is dense in $L^p(\mathbb{R}^m)$ for every $p \in [1, \infty)$.

□

Hence a class of RBF networks with the same smoothing factor in each kernel node is broad enough for universal approximation.

Furthermore it can be shown that RBF networks possess a best representation property in addition to their universal approximation ability. This was proven by Poggio & Girosi (1990) as follows. Let $S = \{(x_i, y_i) \in \mathbb{R}^n \times \mathbb{R} | i = 1, ..., N\}$ be a set of input/output data that we want to approximate by means of a function $f$. According to the regularization approach in approximation theory the following functional is minimized

$$\min_f H[f] = \sum_{i=1}^{N} (y_i - f(x_i))^2 + \lambda \, ||\mathcal{P} f||^2 \tag{2.8}$$

where $\mathcal{P}$ is a constraint operator (usually a differential operator), $||.||$ is a norm on the function space to which $\mathcal{P}f$ belongs (usually the 2-norm) and $\lambda$ is a positive real number (regularization parameter). For $\lambda = 0$ one has an interpolation problem. Minimization of the functional $H$ leads to a partial differential equation (Euler-Lagrange equation), that can be written as

$$\mathcal{P}^\dagger \mathcal{P} f(x) = \frac{1}{\lambda} \sum_{i=1}^{N} (y_i - f(x))\delta(x - x_i)$$

where $\mathcal{P}^\dagger$ is the adjoint of the differential operator $\mathcal{P}$. The solution can be expressed in terms of the Green's function $G$ satisfying the distributional differential equation

$$\mathcal{P}^\dagger \mathcal{P}\, G(x; y) = \delta(x - y).$$

The optimal solution can be written as

$$f(x) = \sum_{i=1}^{N} c_i\, G(x; x_i). \tag{2.9}$$

The solution $f$ lies in an $N$-dimensional subspace of the space of smooth functions. The basis for this subspace is given by the $N$ functions $G(x; x_i)$. Here $G(x; x_i)$ is the Green's function 'centered' at the point $x_i$ and the points $x_i$ are the centers of the expansion. The coefficients $c_i$ are the solution the following linear system:

$$(G + \lambda I)c = y$$

where $(y)_i = y_i$, $(c)_i = c_i$ and $(G)_{ij} = G(x_i; y_i)$. Because the operator $\mathcal{P}^\dagger \mathcal{P}$ is self-adjoint, its Green's function is symmetric $G(x; y) = G(y; x)$ and the matrix $G$ is symmetric. The matrix $G + \lambda I$ is of full rank (unless $-\lambda$ is equal to one of the eigenvalues).

Now depending on the specific choice of the stabilizer $\mathcal{P}$ in (2.8), one obtains several possible solutions. The results are closely related to spline theory (see Poggio & Girosi (1990)). By taking

$$\|\mathcal{P}^L f\|^2 = \sum_{m=0}^{L} a_m \|O^m f\|^2, \quad \|O^m f\|^2 = \sum_{i_1 \ldots i_m}^{n} \int_{\mathbb{R}^n} [\frac{\partial^m f(x)}{\partial x_{i_1} \ldots x_{i_m}}]^2 dx$$

with $a_m = \sigma^{2m} / (m! 2^m)$, the following Green's function is obtained

$$G(x) = A \exp[-(\|x\|^2 / 2\sigma^2)]$$

where $A$ is a normalization constant. The regularized solution is then a linear superposition of Gaussians centered at the $N$ data points $x_i$. Hence for the specific choice of the stabilizer $\mathcal{P}$ shown above, the RBF network is the optimal solution to the approximation problem (2.8).

## 2.3 Classical paradigms of learning

In this Section we review some basic learning rules for multilayer feedforward neural networks, RBF networks and recurrent neural networks.

### 2.3.1 Backpropagation

Given a training set of input/output data, the original learning rule for multilayer perceptrons is the backpropagation algorithm (see Rumelhart *et al.* (1986)). Consider a neural network with $L$ layers (i.e. $L-1$ hidden layers), described as

$$x_{i,p}^{l} = \sigma(\xi_{i,p}^{l}), \qquad \xi_{i,p}^{l} = \sum_{j} w_{ij}^{l} x_{j,p}^{l-1}, \qquad l = 1, ..., L$$

where $l$ is the layer index, $p$ the pattern index and $\sigma(.)$ the activation function. Given are $P$ input patterns and corresponding desired output patterns. Classically the objective to be minimized is the cost function

$$\min_{w_{ij}^{l}} E = \frac{1}{P} \sum_{p=1}^{P} E_p \qquad E_p = \frac{1}{2} \sum_{i=1}^{N_L} (x_{i,p}^{d} - x_{i,p}^{L})^2 \tag{2.10}$$

where $x_p^d$ is the desired output vector corresponding to the $p$-th input pattern, $E_p$ is the contribution of the $p$-th pattern to the cost function $E$ and $N_l$ denotes the number of neurons at layer $l$. The backpropagation algorithm or generalized delta rule is

$$\left\{ \begin{array}{lcll} \Delta w_{ij}^{l} & = & \eta \, \delta_{i,p}^{l} \, x_{j,p}^{l-1} & \\ \delta_{i,p}^{L} & = & (x_{i,p}^{d} - x_{i,p}^{L}) \, \sigma'(\xi_{i,p}^{L}) & \\ \delta_{i,p}^{l} & = & (\sum_{r=1}^{N_{l+1}} \delta_{r,p}^{l+1} w_{ri}^{l+1}) \, \sigma'(\xi_{i,p}^{l}), & l = 1, ..., L-1 \end{array} \right. \tag{2.11}$$

where $\eta$ is the learning rate and the so-called $\delta$ variables are defined as $\delta_{i,p}^{l} = \frac{\partial E_p}{\partial \xi_{i,p}^{l}}$.

The term 'backpropagation' can be understood as follows. Given the current interconnection weights of the neural net, the outputs of the neurons at the different layers are calculated. Secondly the error which is evaluated at the output layer is backpropagated, from the output layer towards the input layer by computing the $\delta$ variables at each layer. Once the $\delta$ variables are known the interconnection weight $w_{ij}^{l}$ are adapted. Backpropagation can be used as an on-line as well as an off-line learning algorithm for minimizing the cost function (2.10). In the on-line mode the weights are updated each time after presenting a given input pattern to the network. In off-line learning the weights are updated each epoch, which means after running once through the $P$ patterns of the training set. A more refined version of the learning rule (2.11), takes into account a momentum term in the adaptation law

$$\Delta w_{ij}^{l}(k+1) = \eta \, \delta_{i,p}^{l} \, x_{j,p}^{l-1} + \alpha \, \Delta w_{ij}^{l}(k). \tag{2.12}$$

In fact the backpropagation algorithm is just a way to express the gradient of the cost function (2.10). From the viewpoint of optimization theory the rule (2.11) is a steepest descent algorithm, which is known to converge slowly. Accelerated backpropagation algorithms are e.g. SuperSAB (Tollenaere (1990)) and Quickprop (Fahlmann (1988)). Fast learning algorithms will be discussed in Section 3.4.

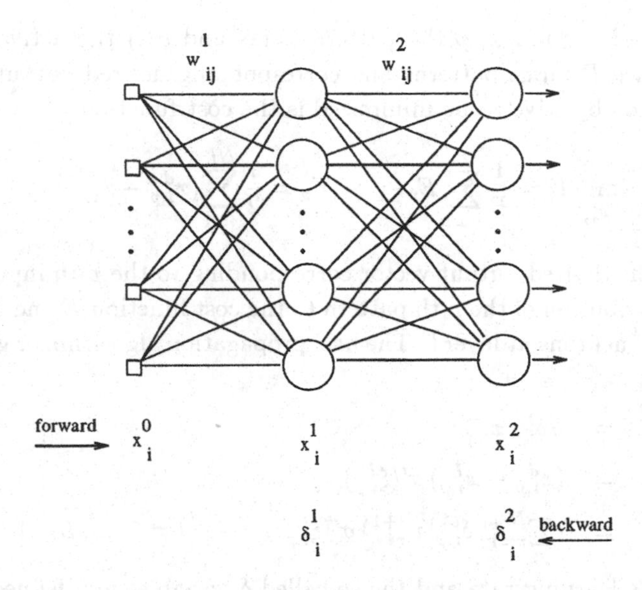


Figure 2.4: *The Backpropagation algorithm: given the current interconnection weights of the neural net, the outputs of the neurons at the different layers are calculated. Secondly the error which is evaluated at the output layer is backpropagated, from the output layer towards the input layer by computing the $\delta$ variables at each layer. Once the $\delta$ variables are known the interconnection weight $w_{ij}^l$ are adapted. In batch mode the adaptation of the weights is done each epoch after presenting the complete training set of input patterns. In on-line mode the adaptation is done each time after presenting a pattern of the training set.*

In the case of supervised learning for recurrent neural networks the calculation of the gradient of the cost function becomes more complicated. Narendra & Parthasarathy (1990)(1991) showed that the static backpropagation algorithm has to be replaced by so-called *dynamic* backpropagation. In order to illustrate this procedure, let us e.g. consider a recurrent neural network of the form

$$\begin{cases} x_{k+1} = f(x_k, u_k; \theta_f), & x_0 \text{ given} \\ y_k = g(x_k, u_k; \theta_g) \end{cases} \tag{2.13}$$

with state vector $x_k \in \mathbb{R}^n$, input vector $u_k \in \mathbb{R}^m$ and output vector $y_k \in \mathbb{R}^l$. The nonlinear mappings $f(.)$ and $g(.)$ are multilayer feedforward neural networks, respectively parametrized by the parameter vectors $\theta_f$ and $\theta_g$. Note that the Hopfield network (2.7) is a special case of (2.13). The training set consists of the input/output data $\{u_k, d_k\}_{k=1}^{k=N}$, where $d_k \in \mathbb{R}^l$ denotes the desired output vector of the recurrent network at time $k$, corresponding to the given input vector $u_k$. The aim is then, e.g., to minimize the cost function

$$\min_{\theta_f, \theta_g} J(\theta_f, \theta_g) = \frac{1}{2N} \sum_{k=1}^{N} (d_k - y_k)^T (d_k - y_k). \tag{2.14}$$

In case one applies a gradient based optimization scheme (many methods will be discussed in Chapter 3 in the context of nonlinear system identification) one needs to know the expressions of $\frac{\partial y_k}{\partial \alpha}$ and $\frac{\partial y_k}{\partial \beta}$ where $\alpha$ and $\beta$ are elements of the parameter vectors $\theta_f$ and $\theta_g$ respectively. This gradient can then be found by a sensitivity method, leading to the following expressions

$$\begin{cases} \frac{\partial x_{k+1}}{\partial \alpha} = \frac{\partial f}{\partial x_k} \cdot \frac{\partial x_k}{\partial \alpha} + \frac{\partial f}{\partial \alpha} \\ \frac{\partial y_k}{\partial \alpha} = \frac{\partial g}{\partial x_k} \cdot \frac{\partial x_k}{\partial \alpha} \\ \frac{\partial y_k}{\partial \beta} = \frac{\partial g}{\partial \beta}. \end{cases} \tag{2.15}$$

This sensitivity model is a dynamical system with state vector $\frac{\partial \hat{x}_k}{\partial \alpha} \in \mathbb{R}^n$ driven by the input vector consisting of $\frac{\partial f}{\partial \alpha} \in \mathbb{R}^n$, $\frac{\partial g}{\partial \beta} \in \mathbb{R}^l$. At the output $\frac{\partial y_k}{\partial \alpha} \in \mathbb{R}^l$, $\frac{\partial y_k}{\partial \beta} \in \mathbb{R}^l$ are generated. Hence the sensitivity model involves the Jacobian matrices $\frac{\partial f}{\partial x_k} \in \mathbb{R}^{n \times n}$ and $\frac{\partial g}{\partial x_k} \in \mathbb{R}^{l \times n}$ and represents the linearized equations of the recurrent neural network around the nominal trajectory and the input. A similar procedure was already applied in the 1960's for the optimization of linear dynamical systems in linear adaptive control (see Narendra & Parthasarathy (1991)).

Hence one can observe that the calculation of gradients of the models (especially dynamic) that contain neural network architectures is rather complicated. In the work of Werbos (1974)(1990) this led to the formulation of a formalism which provides a straightforward manner for calculating the gradients. This is done by considering an ordered set of equations and ordered partial derivatives. Let the set of variables $\{z_i\}$ $(i = 1, .., n)$ describe the variables and unknown parameters of a static neural network or a recurrent neural network through time. Then the following ordered set of equations is taken:

$$\left\{ \begin{array}{rcl} z_2 & = & f_1(z_1) \\ z_3 & = & f_2(z_1, z_2) \\ z_4 & = & f_3(z_1, z_2, z_3) \\ \vdots & = & \vdots \\ z_n & = & f_{n-1}(z_1, z_2, z_3, ..., z_{n-1}) \\ E & = & f_n(z_1, z_2, z_3, ..., z_{n-1}, z_n). \end{array} \right. \tag{2.16}$$

In the last equation of the ordered set the cost function is expressed as a function of the variables and the parameters of the network. An ordered partial derivative is then defined as

$$\frac{\partial^+ z_j}{\partial z_i} = \frac{\partial z_j}{\partial z_i} |_{\{z_1, ..., z_{i-1}\}\text{held constant}} \tag{2.17}$$

The following chain rules for the ordered derivatives hold then

$$\left\{ \begin{array}{rcl} \frac{\partial^+ E}{\partial z_i} & = & \frac{\partial E}{\partial z_i} + \sum_{k>i} \frac{\partial^+ E}{\partial z_k} \frac{\partial z_k}{\partial z_i} \\ \\ \frac{\partial^+ E}{\partial z_i} & = & \frac{\partial E}{\partial z_i} + \sum_{k>i} \frac{\partial E}{\partial z_k} \frac{\partial^+ z_k}{\partial z_i}. \end{array} \right. \tag{2.18}$$

In Werbos (1990) this procedure is called 'backpropagation through time', because the neural network is unfolded in time according to the description (2.16). For example if we consider a dynamical system $x_{k+1} = f(x_k; \alpha)$ and a cost function $E(\alpha)$ is defined, then application of the chain rules gives (see also Piché (1994))

$$\left\{ \begin{array}{rcl} \frac{\partial^+ E}{\partial \alpha} & = & \sum_k \lambda_k^T \frac{\partial x_k}{\partial \alpha} \\ \lambda_k^T & = & \frac{\partial E}{\partial x_k} + \lambda_{k+1}^T \frac{\partial f}{\partial x_k}, \qquad \lambda_N = 0 \end{array} \right. \tag{2.19}$$

and

$$\left\{ \begin{array}{rcl} \frac{\partial^+ E}{\partial \alpha} & = & \sum_k \frac{\partial E}{\partial x_k} z_k \\ z_{k+1} & = & \frac{\partial f}{\partial x_k} z_k + \frac{\partial f}{\partial \alpha}, \qquad z_0 = 0 \end{array} \right. \tag{2.20}$$

where $\lambda_k^T = \frac{\partial^+ E}{\partial x_k}$ and $z_k = \frac{\partial^+ x_k}{\partial \alpha}$. The first chain rule delivers an equation which is running backward in time and is related to the adjoint equation from optimal control theory. The second chain rule gives a sensitivity model like in Narendra's dynamic backpropagation procedure.

We have focused in this Section mainly on methods for computing derivatives in static or dynamic systems containing neural network architectures. However many more aspects are important for successful learning such as, e.g., the choice of the number of hidden neurons, the use of fast local optimization schemes or the use of global optimization methods, model validation and others. All this is discussed more extensively in Chapter 3 in the context of nonlinear system identification using neural networks.

### 2.3.2 RBF networks

In learning of feedforward or recurrent networks, containing sigmoidal activation functions, one ends up with nonlinear optimization problems that exhibit many local optima. This problem is more or less avoided in RBF networks. The idea there is to let the centers of the Gaussians cover the input space at the places where a good approximation is needed. Once the centers are fixed the parametrization is in the output weights of the network only. Moreover this parametrization is linear such that linear regression can be applied in order to learn the output weights.

In Poggio & Girosi (1990) this principle is theoretically motivated as follows. In Section 2.2 we have already explained the best representation property of RBF networks. It was shown that the optimal solution to (2.8) is

$$f(x) = \sum_{i=1}^{N} c_i \, G(\|x - x_i\|).$$

In order to obtain a learning algorithm of some practical relevance, a fewer number of hidden neurons $n_h < N$ is imposed instead of the number of data points $N$. Furthermore a weighted norm $\|.\|_W$ is taken. Instead of (2.9) the following solution is then proposed

$$f^*(x) = \sum_{i=1}^{n_h} c_i \phi_i(x), \tag{2.21}$$

with $\{\phi_i\}_{i=1}^{n_h}$ linearly independent functions. The optimal solution is then obtained by setting $\frac{\partial H[f^*]}{\partial c_i} = 0$ $(i = 1, ..., n_h)$ and is equal to

$$f^*(x) = \sum_{\alpha=1}^{n_h} c_\alpha G(x; t_\alpha) \tag{2.22}$$

with $t_\alpha$ the centers of $G(.)$. Remark that in the special case $n_h = N$ we have $\{t_\alpha\}_{\alpha=1}^N = \{x_i\}_{i=1}^N$. For a weighted norm the goal is then to minimize

$$\min_{c_\alpha, t_\alpha, W} H = \sum_{i=1}^{N} (y_i - \sum_{\alpha=1}^{n_h} c_\alpha G(\|x - t_\alpha\|_W^2))^2. \tag{2.23}$$

The gradient of this cost function is

$$\left\{ \begin{array}{lcl} \frac{\partial H}{\partial c_\alpha} & = & -2 \sum_{i=1}^{N} \Delta_i G(\|x - t_\alpha\|_W^2) \\ \frac{\partial H}{\partial t_\alpha} & = & 4 c_\alpha \sum_{i=1}^{N} \Delta_i G'(\|x - t_\alpha\|_W^2) W^T W (x_i - t_\alpha) \\ \frac{\partial H}{\partial W} & = & -4W \sum_{\alpha=1}^{n_h} c_\alpha \sum_{i=1}^{N} \Delta_i G'(\|x - t_\alpha\|_W^2) Q_{i,\alpha} \end{array} \right. \tag{2.24}$$

with $\Delta_i = y_i - f^*(x_i)$ and $Q_{i,\alpha} = (x_i - t_\alpha)(x_i - t_\alpha)^T$. Here $\sum_{i=1}^{N} Q_{i,\alpha}$ is an estimate for the correlation matrix of all the samples $x_i$ relative to the centers $t_\alpha$. The expressions (2.24) have the following meaning. The expression for $\frac{\partial H}{\partial c_\alpha}$ means that the correction is equal to the sum over the examples of the products between the error on that example and the activity of the hidden unit representing the example with its center. For fixed centers and unweighted norm the expression for $\frac{\partial H}{\partial t_\alpha}$ means clustering because the optimal centers will be equal to a weighted sum of the data points

$$t_\alpha = \frac{\sum_{i=1}^{N} P_i^\alpha x_i}{\sum_{i=1}^{N} P_i^\alpha}, \qquad \alpha = 1, ..., n_h \tag{2.25}$$

with $P_i^\alpha = \Delta_i G'(\|x_i - t_\alpha\|^2)$. Finally the expression for $\frac{\partial H}{\partial W}$ is related to finding an optimal metric.

Hence although the optimization problem (2.23) is non-convex, the following procedure is often used and will yield an acceptable suboptimal solution:

1. Do an unsupervised learning step on the data belonging to the input space of the RBF network. This step consists of placing $n_h$ cluster centers, e.g. by means of the well known K-means clustering algorithm (see Tou & Gonzalez (1974)). This is motivated by the expression for $\frac{\partial H}{\partial t_\alpha}$ in (2.24).

2. Apply linear regression in order to find the output weights.

This algorithm was described in Chen & Billings (1992) in the context of nonlinear system identification. Hence although this procedure is suboptimal, it avoids the problem of local optima because of the linear regression for finding the output weights.

## 2.4 Conclusion

In this Chapter we have discussed multilayer perceptrons and radial basis function networks. Important is that neural networks are universal approximators. This means they can be used for parametrizing any continuous nonlinear function, which highly motivates their use for modelling and control purposes. Moreover, a theorem by Barron states that multilayer perceptrons avoid the curse of dimensionality under certain conditions. The universal approximation theorems are not constructive in the sense that they are existence theorems. Hence there is a need for developing learning algorithms, in order to train the network from a given set of examples. The backpropagation algorithm is the classical learning paradigm for gradient based optimization. The computation of the gradient in recurrent networks is more difficult than in feedforward networks. According to Narendra's dynamic backpropagation the gradient is generated by a sensitivity model, which is in itself also a dynamical system. Another paradigm is backpropagation through time, introduced by Werbos. In that case one makes use of ordered derivatives for a network that is unfolded in time.

Radial basis function networks on the other hand possess besides their universal approximation ability also a best representation property. The theory of RBF networks is closely related to the theory of splines. A suboptimal learning algorithm consists of first placing the centers of the Gaussians and secondly applying linear regression in order to find the output weights.

# Chapter 3

# Nonlinear system identification using neural networks

In this Chapter we treat the problem of nonlinear system identification using neural networks. Model structures and their parametrization by multilayer perceptrons are discussed, together with learning algorithms, practical aspects and examples. The Chapter is organized as follows. In Section 3.1 we review model structures such as NARX, NARMAX and nonlinear state space models. In Section 3.2 parametrizations of these models by multilayer neural nets are made. Neural state space models are introduced. In Section 3.3 classical as well as advanced on- and off-line learning algorithms are presented and their relation with nonlinear optimization theory is explained in Section 3.4. Section 3.5 concerns practical aspects of model validation, model complexity and aspects of pruning and regularization. In Section 3.6 neural network models are interpreted as uncertain linear systems. Finally in Section 3.7 simulated and real life examples are presented on nonlinear system identification using feedforward as well as recurrent type of neural networks. New contributions are made in Sections 3.2.2, 3.2.3, 3.3.2, 3.6 and 3.7.

## 3.1 From linear to nonlinear dynamical models

In *linear* system identification many models can be considered as special cases of the following generalized discrete time model structure

$$A(q)y_k = \frac{B(q)}{F(q)}u_k + \frac{C(q)}{D(q)}e_k \ , \tag{3.1}$$

where $A(q)$,$B(q)$,$C(q)$,$D(q)$,$F(q)$ are (matrix) polynomials in the delay operator $q$ ($qu_k = u_{k-1}$) (see Ljung (1987)). The input and output signal is $u_k \in \mathbb{R}^m$ and $y_k \in \mathbb{R}^l$ respectively. The signal $e_k \in \mathbb{R}^r$ is a white noise exogenous input. Commonly used models can be considered as special cases of (3.1) and are presented in Table 3.1.

| Used polynomials | Name of model structure |
|---|---|
| B | FIR (Finite Impulse Response) |
| AB | ARX (AutoRegressive with eXternal input) |
| AC | ARMA (AutoRegressive Moving Average) |
| ABC | ARMAX (ARMA with eXternal input) |
| BF | OE (Output Error) |
| BFCD | BJ (Box-Jenkins) |

Table 3.1: *Some special cases of the generalized linear model structure* $A(q)y_k = [B(q)/F(q)]u_k + [C(q)/D(q)]e_k$.

The step towards nonlinear modelling is made as follows. The ARX model is extended to a nonlinear ARX model (NARX) as

$$y_k = f(y_{k-1}, y_{k-2}, ..., y_{k-n_y}, u_{k-1}, u_{k-2}, ..., u_{k-n_u}) + e_k \tag{3.2}$$

and the ARMAX model to a nonlinear ARMAX model (NARMAX) as

$$y_k = f(y_{k-1}, y_{k-2}, ..., y_{k-n_y}, u_{k-1}, u_{k-2}, ..., u_{k-n_u}, e_{k-1}, e_{k-2}, ..., e_{k-n_e}) + e_k \tag{3.3}$$

with output vector $y_k \in \mathbb{R}^l$, input vector $u_k \in \mathbb{R}^m$, exogenous input $e_k \in \mathbb{R}^l$ and $n_y, n_u, n_e$ are the lags for the output, input and noise signal respectively. The models (3.1)-(3.3) are all input/output models.

On the other hand one considers also nonlinear *state space* models

$$\left\{ \begin{array}{rcl} x_{k+1} & = & f(x_k, u_k, \varphi_k) \\ y_k & = & g(x_k, u_k, \psi_k) \end{array} \right. \tag{3.4}$$

with $x_k \in \mathbb{R}^n$ the state vector and $\varphi_k \in \mathbb{R}^n$, $\psi_k \in \mathbb{R}^l$ respectively process noise and measurement noise, assumed to be zero mean white Gaussian. A special case of (3.4) is the linear system in state space form

$$\begin{cases} x_{k+1} &= Ax_k + Bu_k + \varphi_k \\ y_k &= Cx_k + Du_k + \psi_k. \end{cases}$$

In nonlinear system identification using neural networks, parametrizations of the nonlinear mappings in (3.2),(3.4) by means of feedforward neural networks are made. This makes sense because the neural nets are universal approximators as was discussed in Chapter 2. This results in powerful model structures that are able to represent complex nonlinear behaviour such as e.g. chaos, hysteresis or saturation effects or combinations of several nonlinear phenomena. Learning rules for neural network models will result in problems of nonlinear regression which will be discussed in the sequel of this Chapter. On the other hand it is well-known that parametrizations of the nonlinear mapping in (3.2) by polynomial expansions result in linear regression, which avoids the problem of many local optima. So why would one prefer parametrizations by neural networks over polynomial expansions ? One argument is in the work of Barron (1993) which shows that parametrizations by neural networks are surprisingly advantageous over polynomial expansions in higher dimensional settings, as we explained in Chapter 2 (Section 2.2.1). This certainly motivates the study of parametrizations by feedforward neural networks.

# 3.2 Parametrization by ANNs

## 3.2.1 Input/output models

For the NARX model (3.2) and the NARMAX model (3.3) a parametrization of the nonlinear mapping $f(.)$ is denoted as

$$y_k = f(x_{k-1}; \theta) + e_k \tag{3.5}$$

where $x_{k-1} = [y_{k-1}; ...; y_{k-n_y}; u_{k-1}; ...; u_{k-n_u}]$ in the case of a NARX model and $x_{k-1} = [y_{k-1}; ...; y_{k-n_y}; u_{k-1}; ...; u_{k-n_u}; e_{k-1}; ...; e_{k-n_e}]$ in the case of a NARMAX model. For a parametrization by a multilayer feedforward neural network with one hidden layer

$$y_k = W \tanh(Vx_{k-1} + \beta) + e_k \tag{3.6}$$

the parameter vector $\theta$ contains the elements of the interconnection matrices $W \in \mathbb{R}^{l \times n_h}$, $V \in \mathbb{R}^{l \times n_x}$ and the bias vector $\beta \in \mathbb{R}^{n_h}$. Here $n_x$ denotes the size of the input vector of the neural network which is equal to $ln_y + mn_u$ for NARX and $l(n_y + n_e) + mn_u$ for NARMAX. Such parametrizations were proposed in Billings *et al.* (1992), Chen *et al.* (1990a,b), Chen & Billings (1992).

In Narendra & Parthasarathy (1990) a difference is made between the following two modes

1. *Series-parallel mode:*
   In this case the measured outputs are used in the input argument of the neural network model, e.g. in the deterministic case

$$\hat{y}_k = f(y_{k-1}, ..., y_{k-n_y}, u_{k-1}, ..., u_{k-n_u}; \theta) \tag{3.7}$$

   and $\hat{y}_k$ is the estimated output. Because there is no recurrence in the equation (3.7), this corresponds to a feedforward neural network. One has a static nonlinear mapping from the input to the output space of the neural network.

2. *Parallel mode:*
   Here the estimated outputs are used in the input argument of the neural network model

$$\hat{y}_k = f(\hat{y}_{k-1}, ..., \hat{y}_{k-n_y}, u_{k-1}, ..., u_{k-n_u}; \theta). \tag{3.8}$$

   Because of the recurrence of $\hat{y}_k$ in the equation, this corresponds to a recurrent neural network.

Hence parametrizing the nonlinear mapping $f(.)$ in (3.2) by a feedforward neural network may either lead to an overall feedforward or a recurrent neural network architecture. The difference between series-parallel models and parallel models is quite essential because the learning algorithms for recurrent neural networks are more complicated than for feedforward neural networks, as we already explained in Chapter 2. Furthermore let us remark that for time series prediction the models (3.7)(3.8) are used as autonomous systems without external input $u_k$, yielding

$$\hat{y}_k = f(y_{k-1}, ..., y_{k-n_y}; \theta) \tag{3.9}$$

in series-parallel mode and

$$\hat{y}_k = f(\hat{y}_{k-1}, ..., \hat{y}_{k-n_y}; \theta) \tag{3.10}$$

in parallel mode.

### 3.2.2 Neural state space models

Parametrizations of nonlinear state space models by feedforward neural networks are discussed in Suykens (1994b). In the case of deterministic identification ($\varphi_k = 0$, $\psi_k = 0$ in (3.4)) one has the following simulation model as predictor

$$\begin{cases} \hat{x}_{k+1} &= f(\hat{x}_k, u_k, 0; \theta) \ ; \ \hat{x}_0 = x_0 \text{ given} \\ \hat{y}_k &= g(\hat{x}_k, u_k, 0; \theta). \end{cases} \tag{3.11}$$

For the stochastic case ($\varphi_k \neq 0$, $\psi_k \neq 0$ in (3.4)) the following predictor can be considered

$$\begin{cases} \hat{x}_{k+1} &= f(\hat{x}_k, u_k, 0; \theta) + K(\theta)\epsilon_k \ ; \ \hat{x}_0 = x_0 \text{ given} \\ \hat{y}_k &= g(\hat{x}_k, u_k, 0; \theta). \end{cases} \tag{3.12}$$

Such a form was proposed in Goodwin & Sin (1984) in the context of adaptive filtering. Here $\epsilon_k = y_k - \hat{y}_k$ denotes the prediction error. A direct parametrization of the steady state Kalman gain $K$ is taken instead of indirectly through the Riccati equation as a result of extended Kalman filtering. This aspect will be explained in Section 3.3 on extended Kalman filtering. Remark that for linear systems

$$\begin{cases} \hat{x}_{k+1} &= A\hat{x}_k + Bu_k + K\epsilon_k \\ y_k &= C\hat{x}_k + Du_k + \epsilon_k \end{cases} \tag{3.13}$$

with $E\{\epsilon_k \epsilon_s^T\} = \Lambda \delta_{ks}$ and $\hat{x}_0 = 0$, the predictor (3.12) reduces to the well-known Kalman filter (Ljung (1979)(1987), Anderson & Moore (1979)) and (3.13) is a model in innovations form with white noise innovations input $\epsilon_k$.

Taking parametrizations by multilayer feedforward neural networks for the nonlinear mappings $f(.)$ and $g(.)$ in (3.11) and (3.12) one obtains

$$\begin{cases} \hat{x}_{k+1} &= W_{AB} \tanh(V_A \hat{x}_k + V_B u_k + \beta_{AB}) + K\epsilon_k \ ; \ \hat{x}_0 = x_0 \\ \hat{y}_k &= W_{CD} \tanh(V_C \hat{x}_k + V_D u_k + \beta_{CD}). \end{cases} \tag{3.14}$$

For deterministic identification one has $K = 0$. The predictor (3.14) is called a *neural state space model* in Suykens *et al.* (1994b). Like input/output models in parallel mode, neural state space models are <u>recurrent</u> neural network architectures. The architecture is shown in Figure 3.1. The dimensions of the interconnection matrices and bias vectors are $W_{AB} \in \mathbb{R}^{n \times n_{hx}}$, $V_A \in \mathbb{R}^{n_{hx} \times n}$, $V_B \in \mathbb{R}^{n_{hx} \times m}$, $\beta_{AB} \in \mathbb{R}^{n_{hx}}$, $W_{CD} \in \mathbb{R}^{l \times n_{hy}}$, $V_C \in \mathbb{R}^{n_{hy} \times n}$, $V_D \in \mathbb{R}^{n_{hy} \times m}$, $\beta_{CD} \in \mathbb{R}^{n_{hy}}$, $K \in \mathbb{R}^{n \times l}$, where the number of hidden neurons of the neural network architectures are $n_{hx}$ and $n_{hy}$. The reason for the specific choice of the indices in the notation of the interconnection matrices will be clear after reading Section 3.6 and is due to the fact that the predictor (3.14) can be written as an uncertain linear system with $A$, $B$, $C$, $D$ matrices depending on $\hat{x}_k$

and $u_k$. The indices of the interconnection matrices indicate to which of these matrices they are related.

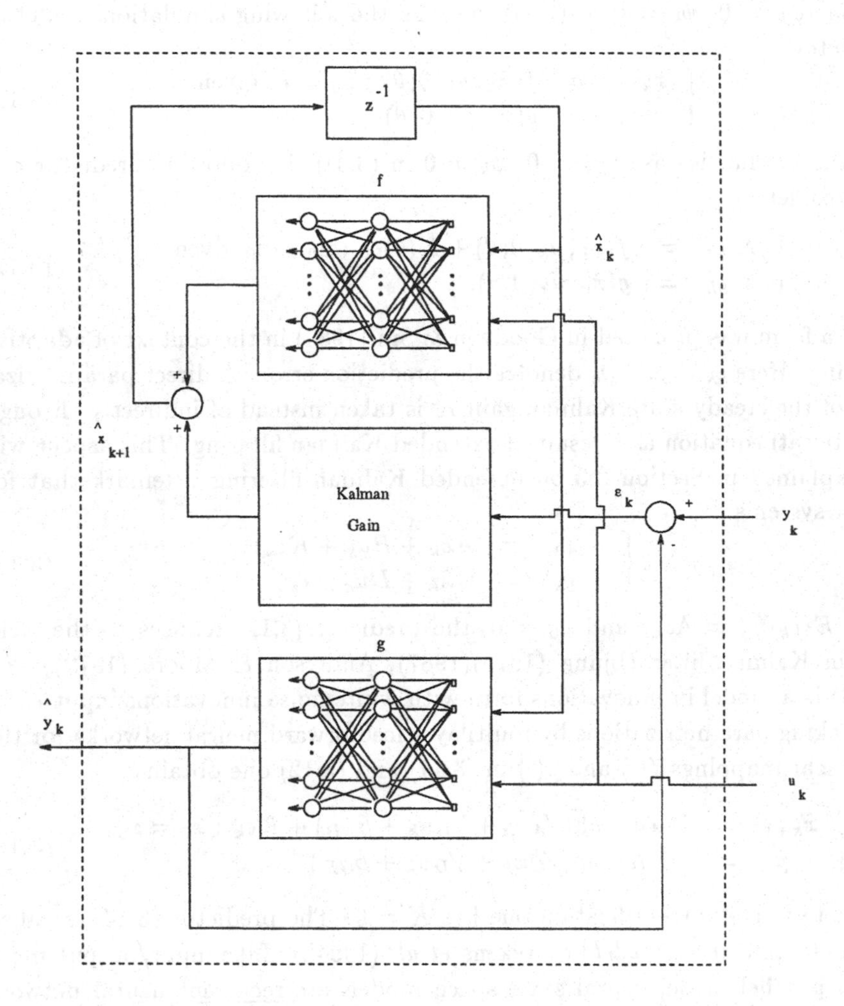


Figure 3.1: *This figure shows a neural state space model. The predictor takes into account process noise and measurement noise, and makes use of a direct parametrization of a steady state Kalman gain. The nonlinear functions $f$ and $g$ are parametrized by multilayer perceptrons.*

### 3.2.3 Identifiability

Identifiability deals with the problem of uniqueness of the weights of the network and is related to the problem whether two networks with different parameter vectors can produce identical input/output behaviour. Work on this topic is done by Sussmann (1992) on feedforward networks and by Albertini & Sontag (1993), Sontag (1993) on recurrent networks. The implication of these Theorems on identifiability of neural state space models will be addressed here.

**Theorem 3.1 [Sussmann].** For feedforward neural nets with a single hidden layer, a single output node and tanh(.) as activation function, the neural network is uniquely determined by its input-output map, up to an obvious finite group of symmetries (permutations of the hidden nodes and changes of the sign of all the weights associated to a particular hidden node), provided that the net is irreducible, i.e. that there does not exists an inner node that makes a zero contribution to the output, and there is no pair of hidden nodes that could be collapsed to a single node without altering the input-output map.

□

The following theorem will be related to continuous time recurrent neural networks of the following form (see Albertini & Sontag (1993))

$$\left\{ \begin{array}{rcl} \dot{x} & = & \sigma(Ax + Bu) \\ y & = & Cx \end{array} \right. \tag{3.15}$$

with matrices $A \in \mathbb{R}^{n \times n}$, $B \in \mathbb{R}^{n \times m}$, $C \in \mathbb{R}^{l \times n}$.

**Definition 3.1 [Sign-permutation, permutation equivalent].** Let $\Sigma = \Sigma(A, B, C, \sigma)$ denote the system (3.15) and $\mathcal{S}(n, m, l)$ the set of all recurrent neural nets $\Sigma$ with fixed $n, m, l$. Two networks $\Sigma$ and $\hat{\Sigma}$ in $\mathcal{S}$ are called sign-permutation equivalent if there exists a nonsingular matrix $T$ such that $\hat{A} = T^{-1}AT$, $\hat{B} = T^{-1}B$, $\hat{C} = CT$ and $T$ has the special form $T = PJ$ where $P$ is a permutation matrix and $J = \text{diag}\{\lambda_1, ..., \lambda_n\}$, with $\lambda_i = \pm 1$ for all $i$. The networks $\Sigma$ and $\hat{\Sigma}$ are called permutation equivalent if the above holds for $J = I$.

Furthermore a subset $\mathcal{C}$ of $\mathcal{S}(n, m, l)$ is called generic if there exists a nonempty subset $\mathcal{G} \subseteq \mathbb{R}^{n^2+nm+nl}$, whose complement is the set of zeros of a finite number of polynomials in $n^2 + nm + nl$ variables, so that $\Sigma$ is in $\mathcal{C}$ if and only if its defining triple $(A, B, C)$ is in $\mathcal{G}$. The following theorem then holds.

**Theorem 3.2 [Albertini & Sontag].** Let the activation function $\sigma$ in (3.15) be odd ($\sigma(-x) = -\sigma(x)$) and satisfy the following assumptions: $\sigma(0) = 0$, $\sigma'(0) \neq 0$, $\sigma''(0) = 0$ and $\sigma^{(q)}(0) \neq 0$ for some $q > 2$ (e.g. tanh(.)). Then for each $n, m, l$ there exists a generic subset $\tilde{\mathcal{S}}(n,m,l) \subseteq \mathcal{S}(n,m,l)$ such that, for any two systems $\Sigma \in \tilde{\mathcal{S}}(n,m,l)$ and $\hat{\Sigma} \in \tilde{\mathcal{S}}(\hat{n},m,l)$, it holds that the networks $\Sigma$ and $\hat{\Sigma}$ are input-output equivalent if and only if $\Sigma$ and $\hat{\Sigma}$ are sign-permutation equivalent.

□

Similar results hold in discrete time (see Sontag (1993)). Hence the recurrent neural network (3.15) is uniquely identifiable from input/output measurements. Under weak genericity assumptions, the following is true. Assume given two networks, with identical activation function. Then if the two networks behave the same as black boxes then they must have the same number of neurons and the same weights, up to sign reversals. Another important implication of the results in Albertini & Sontag (1993) is that dimensionality reduction of the parameter space is impossible for the network (3.15). This is in contrast to linear systems, where canonical forms with a minimal number of parameters are chosen (see e.g. Ljung (1987)). The results of Theorem 3.2 are also closely related to nonlinear realization theory and the geometric theory of nonlinear systems (Isidori (1985)): the systematic use of Lie derivatives is also central in the proof of Theorem 3.2.

**Corollary 3.1.** The following two feedforward networks are input/output equivalent:

$$y = \tanh(x) \tag{3.16}$$

and

$$y = T\tanh(T^{-1}x) \tag{3.17}$$

where $T = PJ$, with $P$ a permutation matrix and $J = \text{diag}\{\pm 1\}$. Starting from this equivalence it is straightforward to understand the result of Theorem 3.2. Indeed starting from a recurrent neural network

$$\begin{cases} z_{k+1} &= \sigma(Az_k + Bu_k) \\ y_k &= Cz_k \end{cases} \tag{3.18}$$

and taking into account the equivalence between (3.16) and (3.17), one obtains

$$\begin{cases} z_{k+1} &= T\sigma(T^{-1}Az_k + T^{-1}Bu_k) \\ y_k &= Cz_k \end{cases} \tag{3.19}$$

Putting $z_k = Tx_k$ with $T$ square and nonsingular, this gives

$$\begin{cases} Tx_{k+1} &= T\sigma(T^{-1}ATx_k + T^{-1}Bu_k) \\ y_k &= CTx_k \end{cases} \tag{3.20}$$

and finally

$$\begin{cases} x_{k+1} &= \sigma(\hat{A}x_k + \hat{B}u_k) \\ y_k &= \hat{C}x_k \end{cases} \tag{3.21}$$

with $\hat{A} = T^{-1}AT$, $\hat{B} = T^{-1}B$, $\hat{C} = CT$, which illustrates the condition of sign-equivalence in Theorem 3.2.

**Corollary 3.2.** Following the same principle as in Corollary 3.1 we have the following input/output equivalent forms for the neural state space model (3.14)

$$\begin{cases} \hat{z}_{k+1} &= W_{AB}\tanh(V_A\hat{z}_k + V_Bu_k + \beta_{AB}) + K\epsilon_k \\ \hat{y}_k &= W_{CD}\tanh(V_C\hat{z}_k + V_Du_k + \beta_{CD}). \end{cases} \tag{3.22}$$

Using the equivalence between (3.16) and (3.17), one obtains

$$\begin{cases} \hat{z}_{k+1} &= W_{AB}T_1\tanh(T_1^{-1}V_A\hat{z}_k + T_1^{-1}V_Bu_k + T_1^{-1}\beta_{AB}) + K\epsilon_k \\ \hat{y}_k &= W_{CD}T_2\tanh(T_2^{-1}V_C\hat{z}_k + T_2^{-1}V_Du_k + T_2^{-1}\beta_{CD}) \end{cases} \tag{3.23}$$

with $T_i = P_iJ_i$ $(i = 1, 2)$, $P_i$ a permutation matrix and $J_i = \mathrm{diag}\{\pm 1\}$ and $T_1 \in \mathbb{R}^{n_{hx} \times n_{hx}}$, $T_2 \in \mathbb{R}^{n_{hy} \times n_{hy}}$. Then putting $z_k = Sx_k$ with $S \in \mathbb{R}^{n \times n}$ and of full rank, one has

$$\begin{cases} \hat{x}_{k+1} &= \hat{W}_{AB}\tanh(\hat{V}_A\hat{x}_k + \hat{V}_Bu_k + \hat{\beta}_{AB}) + \hat{K}\epsilon_k \\ \hat{y}_k &= \hat{W}_{CD}\tanh(\hat{V}_C\hat{x}_k + \hat{V}_Du_k + \hat{\beta}_{CD}) \end{cases} \tag{3.24}$$

with $\hat{W}_{AB} = S^{-1}W_{AB}T_1$, $\hat{V}_A = T_1^{-1}V_AS$, $\hat{V}_B = T_1^{-1}V_B$, $\hat{\beta}_{AB} = T_1^{-1}\beta_{AB}$, $\hat{K} = S^{-1}K$, $\hat{W}_{CD} = W_{CD}T_2$, $\hat{V}_C = T_2^{-1}V_CS$, $\hat{V}_D = T_2^{-1}V_D$, $\hat{\beta}_{CD} = T_2^{-1}\beta_{CD}$. Although we do not have a formal proof, this suggests that the representation of the neural state space model is unique up to a similarity transformation and sign reversals. Remark that for linear state space models this is only up to a similarity transformation.

## 3.3 Learning algorithms

In this Section we discuss learning algorithms for input/output models in series-parallel and parallel mode and for neural state space models. First an overview is given of learning rules for input/output models in series-parallel mode which is a feedforward architecture. Secondly learning for input/output models in parallel mode and neural state space models, which are both recurrent network architectures, is treated.

### 3.3.1 Feedforward network related models

#### 3.3.1.1 Backpropagation algorithm

The application of the generalized delta rule to the input/output NARX model (3.6) in series-parallel mode is straightforward. Given a measured set of I/O data $\{u_k, y_k\}_{k=1}^{k=N_{tot}}$, the data set is divided into a training set of $N$ data for the identification. The remaining data belong to the test set, which is needed in order to validate the identified model. The training set for supervised learning consists then of the input patterns $\{x_{k-1}\}_{k=k_0}^{k=k_f}$ with $k_0 = \max\{n_y, n_u\} + 1$, $k_f = k_0 + N - 1$ and the corresponding output patterns $\{y_k\}_{k=k_0}^{k=k_f}$. The objective to be minimized is

$$\min_{w_{ij}^l} E = \frac{1}{N} \sum_{k=k_0}^{k_f} E_k \qquad E_k = \frac{1}{2} \sum_{i=1}^{N_L} (x_{i,k}^d - x_{i,k}^L)^2 \tag{3.25}$$

where the pattern index becomes the time index $k$. Here $x^d$ is the desired output and $x^L$ the output of the neural network. Taking the one hidden layer neural network (2.5) ($L = 2$) with linear activation function for the output layer, the generalized delta rule (2.11) becomes

$$\begin{cases} \Delta w_{ij}^2 &= \eta\, \delta_{i,k}^2\, x_{j,k}^1 \\ \Delta w_{ij}^1 &= \eta\, \delta_{i,k}^1\, x_{j,k}^0 \\ \delta_{i,k}^2 &= (x_{i,k}^d - x_{i,k}^2) \\ \delta_{i,k}^1 &= (\sum_{r=1}^{N_2} \delta_{r,k}^2 w_{ri}^2)\, \sigma'(\xi_{i,k}^1) \end{cases}$$

where $w_{ij}^1$ and $w_{ij}^2$ denote the weights of the hidden layer and the output layer respectively. The boundary conditions for the multilayer feedforward neural network architecture are

$$x_k^2 = \hat{y}_k, \qquad x_k^d = y_k, \qquad x_k^0 = x_{k-1}. \tag{3.26}$$

The algorithm can be applied on- or off-line and with or without momentum term as was explained in Chapter 2.

#### 3.3.1.2 Prediction error algorithms

The prediction error algorithm is a standard method in system identification (see e.g. Ljung (1987), Goodwin & Sin (1984)). Its application to nonlinear system identification using neural networks is reported in the work of Chen *et al.* (1990a,b)(1992), Billings *et al.* (1992). Some basic aspects of the latter will be reviewed here.

Let us take again the NARX or NARMAX model in series-parallel mode

$$y_k = W \tanh(V x_{k-1} + \beta) + e_k$$

or shortly

$$y_k = \hat{f}(x_{k-1}; \theta) + e_k$$

with $\theta \in \mathbb{R}^{n_\theta}$. Given the N data, in prediction error algorithms, one then aims at minimizing a cost function of the type

$$J(\theta) = \frac{1}{2N} \sum_{k=k_0}^{k_f} l(\epsilon_k(\theta)) \tag{3.27}$$

for off-line identification, where $\epsilon_k(\theta) = y_k - \hat{y}_k(\theta)$ is the prediction error and $l(\epsilon_k) = \epsilon_k^T \Lambda^{-1} \epsilon_k$, a positive function, having a positive definite symmetric matrix $\Lambda$ of dimension $l \times l$. Let the gradient of the output of the neural network with respect to $\theta$ be denoted as

$$\Psi_k(\theta) = [\frac{d\hat{f}(\theta)}{d\theta}]^T = \hat{g}(x_{k-1}; \theta). \tag{3.28}$$

Hence $\Psi_k \in \mathbb{R}^{n_\theta \times l}$ and

$$\left\{ \begin{array}{lcl} \frac{\partial f^i}{\partial w_{jr}} & = & \delta_j^i \tanh(v_r^T x_{k-1} + \beta_r) \\ \frac{\partial f^i}{\partial v_{rs}} & = & w_{ir}\,(1 - \tanh^2(v_r^T x_{k-1} + \beta_r))\, x_s \\ \frac{\partial f^i}{\partial \beta_r} & = & w_{ir}\,(1 - \tanh^2(v_r^T x_{k-1} + \beta_r)) \end{array} \right. \tag{3.29}$$

with indices $r = 1, ..., n_h$ and $s = 1, ..., n_x$ and $v_r^T$ the $r$-th row of the interconnection matrix $V$. The optimization problem (3.27) is a nonlinear least squares problem which can be solved by means of the Gauss-Newton method

$$\theta_t = \theta_{t-1} + s_t\, \eta(\theta_{t-1}, \delta) \tag{3.30}$$

where $\eta$ is the search direction and $s_t$ is the step size for the current estimate in the iteration scheme and

$$\left\{ \begin{array}{lcl} \eta(\theta, \delta) & = & -[H(\theta, \delta)]^{-1} \nabla J(\theta) \\ \nabla J(\theta) & = & -\frac{1}{N} \sum_{k=k_0}^{k_f} \Psi_k \Lambda^{-1} \epsilon_k \\ H(\theta, \delta) & = & \frac{1}{N} \sum_{k=k_0}^{k_f} \Psi_k \Lambda^{-1} \Psi_k^T + \delta I \end{array} \right. \tag{3.31}$$

Here $\delta$ is a positive scalar and the step size $s_t$ is determined by line search along the direction $\eta$ at each iteration. This algorithm is known to converge at least to a local minimum. Nonlinear least squares problems will be discussed in more detail in Section 3.4 on nonlinear optimization.

Besides off-line (batch) algorithms also recursive methods are available for on-line identification. The following standard form of a recursive prediction error algorithm is

$$\left\{\begin{array}{rcl} R_k &=& R_{k-1} + \gamma_k[\Psi_k \Lambda^{-1} \Psi_k^T + \delta I - R_{k-1}] \\ \hat{\theta}_k &=& \hat{\theta}_{k-1} + \gamma_k R_k^{-1} \Psi_k \Lambda^{-1} \epsilon_k \end{array}\right. \tag{3.32}$$

where $\hat{\theta}_k$ is the estimate of $\theta$ at time $k$ and $\gamma_k$ is the gain at time $k$. The term $R_k^{-1} \Psi_k \Lambda^{-1} \epsilon_k$ can be regarded as an approximation to the Gauss-Newton search direction. It can be shown that $\hat{\theta}_k$ converges with probability one to a local minimum of

$$\overline{J}(\theta) = \lim_{N \to \infty} \frac{1}{2N} \sum_{k=1}^{N} E\{l(\epsilon_k)\} \tag{3.33}$$

where $E\{.\}$ denotes the expectation operator. A more reliable algorithm instead of the standard algorithm (3.32) is a trace adjustment technique, which avoids that matrices explode during iteration

$$\left\{\begin{array}{rcl} \overline{P}_k &=& P_{k-1} - P_{k-1}\Psi_k[\lambda I + \Psi_k^T P_{k-1} \Psi_k]^{-1} \Psi_k^T P_{k-1} \\ P_k &=& \frac{K_0}{\mathrm{trace}[\overline{P}_k]} \overline{P}_k \\ \hat{\theta}_k &=& \hat{\theta}_{k-1} + P_k \Delta_k \\ \Delta_k &=& \gamma_m \Delta_{k-1} + \gamma_g \Psi_k \epsilon_k \end{array}\right. \tag{3.34}$$

where $P_k$ is the covariance matrix estimate for $\hat{\theta}_k$, $\gamma_g$ is the adaptive gain, $\gamma_m$ is a momentum parameter, $\lambda$ is a forgetting factor $(0 < \lambda < 1)$ and $K_0$ a positive constant. Furthermore also parallel algorithms were developed in Chen *et al.* (1990b). The 'full' Hessian matrix is approximated by a block diagonal one, where each block corresponds to the parameters of a particular neuron in the network. In case of neural networks with many parameters this significantly reduces the computational complexity.

#### 3.3.1.3 Extended Kalman filtering

Before applying extended Kalman filtering to neural network models, we will review first some basic facts about extended Kalman filtering (EKF) according to Goodwin & Sin (1984). Basically extended Kalman filtering is one possible way for estimating the state of a nonlinear dynamical system that is corrupted by process and measurement noise. This is done by successive linearization of the nonlinear system around a reference trajectory. The algorithm was first proposed in 1963 by Kopp and Orford and in 1964 by Cox.

Given the nonlinear system

$$\begin{cases} x_{k+1} &= f(x_k, u_k, \varphi_k) \\ y_k &= g(x_k, u_k, \psi_k) \end{cases}$$

with covariance matrices

$$E\{\begin{bmatrix} \varphi_k \\ \psi_k \end{bmatrix} [\varphi_s^T \; \psi_s^T]\} = \begin{bmatrix} Q & S \\ S^T & R \end{bmatrix} \delta_{ks}, \tag{3.35}$$

let $\hat{x}_k$ be an estimate of the state $x_k$ and linearize the nonlinear system around $x_k = \hat{x}_k$, $\varphi_k = 0$, $\psi_k = 0$:

$$\begin{cases} x_{k+1} &\simeq f(\hat{x}_k, u_k, 0) + F_k[x_k - \hat{x}_k] + G_k \varphi_k \\ y_k &\simeq g(\hat{x}_k, u_k, 0) + H_k[x_k - \hat{x}_k] + J_k \psi_k \end{cases} \tag{3.36}$$

where

$$\begin{array}{ll} F_k = \frac{\partial f(x_k, u_k, \varphi_k)}{\partial x_k}\big|_{x_k = \hat{x}_k, \varphi_k = 0} & G_k = \frac{\partial f(x_k, u_k, \varphi_k)}{\partial \varphi_k}\big|_{x_k = \hat{x}_k, \varphi_k = 0} \\ H_k = \frac{\partial g(x_k, u_k, \psi_k)}{\partial x_k}\big|_{x_k = \hat{x}_k, \psi_k = 0} & J_k = \frac{\partial g(x_k, u_k, \psi_k}{\partial \psi_k}\big|_{x_k = \hat{x}_k, \psi_k = 0} \end{array}$$

The extended Kalman filter (EKF) is then the Kalman filter applied to the linearized equations, yielding

$$\begin{cases} \hat{x}_{k+1} &= f(\hat{x}_k, u_k, 0) + K_k[y_k - g(\hat{x}_k, u_k, 0)], \quad \hat{x}_0 \text{ given} \\ K_k &= [F_k \Sigma_k H_k^T + \overline{S}_k][H_k \Sigma_k H_k^T + \overline{R}_k]^{-1} \\ \Sigma_{k+1} &= F_k \Sigma_k F_k^T + \overline{Q}_k - K_k[H_k \Sigma_k H_k^T + \overline{R}_k] K_k^T, \quad \Sigma_0 \text{ given} \end{cases} \tag{3.37}$$

with $K_k$ the Kalman gain and

$$\overline{Q}_k = G_k \, Q \, G_k^T, \quad \overline{S}_k = G_k \, S \, J_k^T, \quad \overline{R}_k = J_k \, R \, J_k^T. \tag{3.38}$$

The EKF is however not an optimal state estimator and its convergence is not guaranteed, but nevertheless it is used in many applications. In fact the EKF can be thought of as a restricted complexity filter which is constrained to have a similar format as the Kalman filter for linear systems. Improvements to the algorithm (3.37) have been proposed, such as iteratively refined linearization (see Goodwin & Sin (1984)).

Let us consider now again the NARX model, parametrized by a neural net (3.6), then the EKF can be applied as follows. For (3.6) the nonlinear system

$$\begin{cases} \theta_{k+1} & = & \theta_k \\ y_k & = & h(\theta_k, v_k) + \psi_k \end{cases} \tag{3.39}$$

is taken. Hence the state of the system is the parameter vector $\theta$, that is to be estimated. The output equation corresponds to the NARX model, which is a static nonlinear mapping with as output the measured output of the plant $y_k$ and as input the vector $v_k = x_{k-1}$. The nonlinear map $h(.)$ in (3.39) corresponds to $f(.)$ in (3.5) and $\psi_k$ is measurement noise. Hence $\varphi_k = 0$, $Q = 0$, $S = 0$ and the matrices in the linearization (3.36) become

$$F_k = I, \quad G_k = 0, \quad H_k = \frac{\partial h}{\partial \theta_k}|_{\theta_k = \hat{\theta}_k}, \quad J_k = I. \tag{3.40}$$

One obtains then for (3.37)

$$\begin{cases} \hat{\theta}_{k+1} & = & \hat{\theta}_k + K_k[y_k - h(\hat{\theta}_k, v_k)] \\ K_k & = & \Sigma_k H_k^T [H_k \Sigma_k H_k^T + R]^{-1} \\ \Sigma_{k+1} & = & \Sigma_k - K_k[H_k \Sigma_k H_k^T + R] K_k^T. \end{cases} \tag{3.41}$$

The state augmentation principle of putting $\theta_{k+1} = \theta_k$ in (3.39) is well-known in adaptive filtering (see e.g. Goodwin & Sin (1984)). The EKF was applied to neural networks e.g. in the work of Singhal & Wu (1989) and Shah *et al.* (1992).

### 3.3.2 Recurrent network related models

#### 3.3.2.1 Dynamic backpropagation

The NARX or NARMAX model in parallel mode and neural state space models are recurrent neural networks. As we have seen in Chapter 2, the calculation of the gradient of the cost function becomes more complicated than for static systems.

A procedure for calculating the gradient for nonlinear dynamical systems is given in Narendra & Parthasarathy (1990)(1991). The following two simple examples from Narendra & Parthasarathy (1991) illustrate the differences between input/output models and state space models. The examples are in continuous time, but similar results are immediately applicable to discrete time systems as well.

**Example 1.** Consider the second order scalar nonlinear differential equation

$$\ddot{y} + F(\alpha, \dot{y}) + y = u \tag{3.42}$$

with initial condition $y(0) = y_{1_0}$, $\dot{y}(0) = y_{2_0}$ and $\alpha$ some scalar parameter to be adjusted. Suppose the goal is to optimize the following performance index of interest

$$J(\alpha) = \frac{1}{T} \int_0^T [y(\tau, \alpha) - y_d(\tau)]^2 d\tau \tag{3.43}$$

where $y_d(t)$ is a reference trajectory. For the calculation of the gradient of $J$ with respect to $\alpha$

$$\frac{\partial J(\alpha)}{\partial \alpha} = \frac{1}{T} \int_0^T 2[y(\tau) - y_d(\tau)] \frac{\partial y(\tau)}{\partial \alpha} d\tau$$

the function $\partial y(t)/\partial \alpha$ is needed. This is obtained by differentiating (3.42) with respect to $\alpha$

$$\ddot{z} + \frac{\partial F}{\partial \dot{y}} \dot{z} + z = -\frac{\partial F}{\partial \alpha} \tag{3.44}$$

where $z = \partial y/\partial \alpha$ and the initial state is $z(0) = \dot{z}(0) = 0$. Hence a linear time-varying model is obtained. The model (3.44) is called a sensitivity model. Also remark that in case (3.42) is a linear model, then the sensitivity model is the linear model itself but with another external input and another initial condition.

**Example 2.** Multivariable systems are easier to handle in state space form. Consider e.g. the nonlinear system with

$$\dot{x}(t) = f[x(t), \alpha, u(t)], \quad x(t_0) = x_0 \tag{3.45}$$

where $x(t) \in \mathbb{R}^n$ and $u(t) \in \mathbb{R}^m$ and $\alpha$ is some scalar parameter. Then by differentiating (3.45) with respect to $\alpha$ the sensitivity model becomes

$$\frac{\partial \dot{x}(t)}{\partial \alpha} = f_x(t) \frac{\partial x(t)}{\partial \alpha} + f_\alpha(t), \quad \frac{\partial x(t_0)}{\partial \alpha} = 0 \tag{3.46}$$

where the Jacobian $f_x(t)$ and $f_\alpha(t)$ are evaluated around the nominal values.

In Suykens *et al.* (1994b) this sensitivity method is applied to neural state space models. Taking the model (3.14)

$$\begin{cases} \hat{x}_{k+1} &= W_{AB} \tanh(V_A \hat{x}_k + V_B u_k + \beta_{AB}) + K\epsilon_k \; ; \; \hat{x}_0 = x_0 \\ \hat{y}_k &= W_{CD} \tanh(V_C \hat{x}_k + V_D u_k + \beta_{CD}) \end{cases}$$

and given $N$ input/output data $Z^N$ a prediction error algorithm aims at minimizing the cost function

$$V_N(\theta, Z^N) = \frac{1}{N}\sum_{k=1}^{N} l(\epsilon_k(\theta)) \tag{3.47}$$

in the unknown parameter vector

$$\theta = [W_{AB}(:); V_A(:); V_B(:); \beta_{AB}; K(:); W_{CD}(:); V_C(:); V_D(:); \beta_{CD}]. \tag{3.48}$$

Narendra's dynamic backpropagation procedure applied to (3.14) then means optimizing by means of a steepest descent method

$$\hat{\theta}_{t+1} = \hat{\theta}_t - \eta \frac{\partial V_N}{\partial \theta} \tag{3.49}$$

where $\eta$ is the learning rate. Other optimization methods with faster convergence such as quasi-Newton and conjugate gradient methods are discussed in Section 3.4. Let us take for $l(\epsilon_k) = \frac{1}{2}\epsilon_k^T \epsilon_k$ and denote the predictor (3.14) as

$$\left\{ \begin{array}{rcl} \hat{x}_{k+1} & = & \Phi(\hat{x}_k, u_k, \epsilon_k; \alpha) \; ; \; \hat{x}_0 = x_0 \text{ given} \\ \hat{y}_k & = & \Psi(\hat{x}_k, u_k; \beta) \end{array} \right. \tag{3.50}$$

where $\alpha, \beta$ are elements of the parameter vector $\theta$. The sensitivity model becomes then

$$\left\{ \begin{array}{rcl} \frac{\partial \hat{x}_{k+1}}{\partial \alpha} & = & \frac{\partial \Phi}{\partial \hat{x}_k} \cdot \frac{\partial \hat{x}_k}{\partial \alpha} + \frac{\partial \Phi}{\partial \alpha} \\ \frac{\partial \hat{y}_k}{\partial \alpha} & = & \frac{\partial \Psi}{\partial \hat{x}_k} \cdot \frac{\partial \hat{x}_k}{\partial \alpha} \\ \frac{\partial \hat{y}_k}{\partial \beta} & = & \frac{\partial \Psi}{\partial \beta} \end{array} \right. \tag{3.51}$$

which is a dynamical model with state vector $\frac{\partial \hat{x}_k}{\partial \alpha} \in \mathbb{R}^n$ driven by the input vector consisting of $\frac{\partial \Phi}{\partial \alpha} \in \mathbb{R}^n$, $\frac{\partial \Psi}{\partial \beta} \in \mathbb{R}^l$ and at the output $\frac{\partial \hat{y}_k}{\partial \alpha} \in \mathbb{R}^l$, $\frac{\partial \hat{y}_k}{\partial \beta} \in \mathbb{R}^l$ are generated. The Jacobians $\frac{\partial \Phi}{\partial \hat{x}_k} \in \mathbb{R}^{n \times n}$ and $\frac{\partial \Psi}{\partial \hat{x}_k} \in \mathbb{R}^{l \times n}$ are evaluated around the nominal trajectory. In order to write down the derivatives, let us take an elementwise notation for (3.14)

$$\left\{ \begin{array}{rcl} \hat{x}^i & := & \sum_j w_{AB}{}^i_j \tanh(\sum_r v_A{}^j_r \, \hat{x}^r + \sum_s v_B{}^j_s \, u^s + \beta_{AB}{}^j) + \sum_j \kappa^i_j \, \epsilon^j \\ \hat{y}^i & = & \sum_j w_{CD}{}^i_j \tanh(\sum_r v_C{}^j_r \, \hat{x}^r + \sum_s v_D{}^j_s \, u^s + \beta_{CD}{}^j), \end{array} \right. \tag{3.52}$$

where $\{.\}^i$ and $\{.\}^i_j$ denote respectively the $i$-th element of a vector and the $ij$-th entry of a matrix. The time index $k$ is omitted after introducing the assignment operator ':='. Defining

$$\begin{array}{rcl} \phi^l & = & \sum_r v_A{}^l_r \, \hat{x}^r + \sum_s v_B{}^l_s \, u^s + \beta^l_{AB} \\ \rho^l & = & \sum_r v_C{}^l_r \, \hat{x}^r + \sum_s v_D{}^l_s \, u^s + \beta^l_{CD} \end{array} \tag{3.53}$$

one obtains the derivatives

$$\frac{\partial \Phi}{\partial \alpha} : \begin{cases} \frac{\partial \Phi^i}{\partial w_{AB}{}_l^j} & = & \delta_j^i \tanh(\phi^l) \\ \frac{\partial \Phi^i}{\partial v_{A}{}_l^j} & = & w_{AB}{}_j^i (1 - \tanh^2(\phi^j)) \hat{x}^l \\ \frac{\partial \Phi^i}{\partial v_{B}{}_l^j} & = & w_{AB}{}_j^i (1 - \tanh^2(\phi^j)) u^l \\ \frac{\partial \Phi^i}{\partial \beta_{AB}{}^j} & = & w_{AB}{}_j^i (1 - \tanh^2(\phi^j)) \\ \frac{\partial \Phi^i}{\partial \kappa_l^j} & = & \delta_j^i \epsilon_l \end{cases} \tag{3.54}$$

$$\frac{\partial \Psi}{\partial \beta} : \begin{cases} \frac{\partial \Psi^i}{\partial w_{CD}{}_l^j} & = & \delta_j^i \tanh(\rho^l) \\ \frac{\partial \Psi^i}{\partial v_{C}{}_l^j} & = & w_{CD}{}_j^i (1 - \tanh^2(\rho^j)) \hat{x}^l \\ \frac{\partial \Psi^i}{\partial v_{D}{}_l^j} & = & w_{CD}{}_j^i (1 - \tanh^2(\rho^j)) u^l \\ \frac{\partial \Psi^i}{\partial \beta_{CD}{}^j} & = & w_{CD}{}_j^i (1 - \tanh^2(\rho^j)) \end{cases}$$

$$\frac{\partial \Phi}{\partial \hat{x}_k} : \quad \frac{\partial \Phi^i}{\partial \hat{x}^r} = \sum_j w_{AB}{}_j^i (1 - \tanh^2(\phi^j))\, v_{A}{}_r^j$$

$$\frac{\partial \Psi}{\partial \hat{x}_k} : \quad \frac{\partial \Psi^i}{\partial \hat{x}^r} = \sum_j w_{CD}{}_j^i (1 - \tanh^2(\rho^j))\, v_{C}{}_r^j .$$

Hence calculating the gradient of the neural state space model requires as many simulations of the sensitivity model as the number of elements in the parameter vector $\theta$. Parallelization of the algorithm at this level is straightforward then, because these simulations can be distributed over the available number of processors.

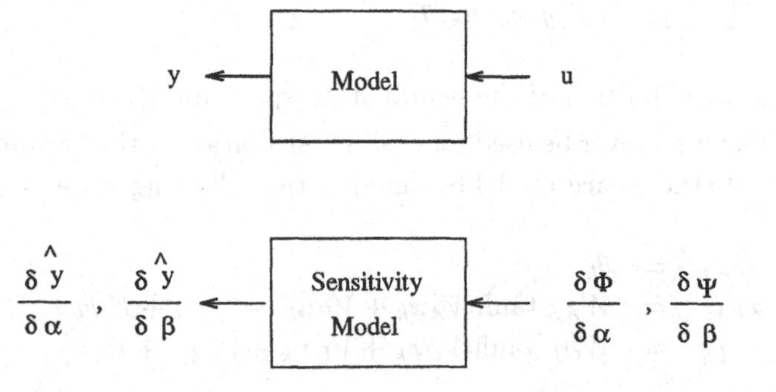


Figure 3.2: *Dynamic backpropagation applied to some nonlinear state space model needs a sensitivity model, which is in itself also a dynamical system, that generates the gradient of the cost function.*

#### 3.3.2.2 Extended Kalman filtering

Suppose the underlying nonlinear system is

$$\left\{ \begin{array}{rcl} x_{k+1} & = & f(x_k, u_k) + \phi_k \\ y_k & = & g(x_k, u_k) + \psi_k \end{array} \right. \tag{3.55}$$

and we aim at using extended Kalman filtering for estimating the state $x_k$ together with parametrizations by neural networks for $f(.)$ and $g(.)$. Then according to (3.37) the EKF gives

$$\left\{ \begin{array}{rcl} \hat{x}_{k+1} & = & f(\hat{x}_k, u_k; \theta) + K_k(\theta)\epsilon_k \ ; \ \hat{x}_0 = x_0 \text{ given} \\ \hat{y}_k & = & g(\hat{x}_k, u_k; \theta) \\ K_k(\theta) & = & [F_k(\theta)\Sigma_k(\theta)H_k(\theta)^T + S(\theta)][H_k(\theta)\Sigma_k(\theta)H_k(\theta)^T + R(\theta)]^{-1} \\ \Sigma_{k+1} & = & F_k(\theta)\Sigma_k F_k(\theta)^T + Q(\theta) - K_k(\theta)[H_k(\theta)\Sigma_k(\theta)H_k(\theta)^T]K_k(\theta)^T, \end{array} \right. \tag{3.56}$$

where $F_k$, $H_k$, $Q$, $S$ and $R$ are defined according to (3.36). Besides $f(.)$ and $g(.)$ the covariance matrices $Q, S, R$ are also parametrized by $\theta$. In case we want to apply a prediction error algorithm for off-line system identification and a gradient based optimization scheme, calculation of the analytic expression for the gradient of the cost function becomes unfeasible because of the complicated dependency through the Riccati equation. This explains the choice of a direct parametrization of the steady state Kalman gain (see e.g. Ljung (1979), Goodwin & Sin (1984)),

$$\left\{ \begin{array}{rcl} \hat{x}_{k+1} & = & f(\hat{x}_k, u_k; \theta) + K(\theta)\epsilon_k \ ; \ \hat{x}_0 = x_0 \text{ given} \\ \hat{y}_k & = & g(\hat{x}_k, u_k; \theta) \end{array} \right. \tag{3.57}$$

which led to the definition of the neural state space model (3.14).

The EKF can however be used for on-line estimation of the parameter vector $\theta$ of the neural state space model by defining the following state augmentation

$$\left\{ \begin{array}{rcl} \theta_{k+1} & = & \theta_k \\ x_{k+1} & = & W_{AB} \tanh(V_A x_k + V_B u_k + \beta_{AB}) + K\psi_k \\ y_k & = & W_{CD} \tanh(V_C x_k + V_D u_k + \beta_{CD}) + \psi_k \end{array} \right. \tag{3.58}$$

where $\theta$ is defined as in (3.48). The augmented state vector is then equal to the vector $[\theta_k; x_k]$. The derivation of the EKF according to (3.37) is then straightforward though complicated.

## 3.4 Elements from nonlinear optimization theory

As we have seen in Chapter 2 for the static and dynamic backpropagation algorithm, many learning problems can be cast as unconstrained optimization problems of the form

$$\min_{x \in \mathbb{R}^n} C(x) \tag{3.59}$$

Let us assume a cost function $C(.)$ that is twice continuously differentiable. The simplest optimization scheme is a steepest descent algorithm

$$x_{t+1} = x_t - \alpha_t \nabla C(x_t) \tag{3.60}$$

where $\alpha_t$ is the step size and $x_t$ the $t$-th iterate. Backpropagation without momentum term is such a steepest descent algorithm. It is well known that such algorithms convergence linearly. Methods with a better convergence rate from nonlinear optimization theory are explained now (see also Fletcher (1987), Gill *et al.* (1981) and Cuthbert (1987)).

Consider a point $x_0$ in $n$-dimensional search space and a Taylor expansion around that point

$$C(x) = C(x_0) + g^T \Delta x + \frac{1}{2} \Delta x^T H \Delta x \tag{3.61}$$

where $\Delta x = x - x_0$ is the step, $g = \nabla C(x_0)$ the gradient at $x_0$ and $H = \nabla^2 C(x_0)$ the Hessian at $x_0$. From this expression follows the optimal Newton step

$$\frac{\partial C}{\partial(\Delta x)} = g + H\,\Delta x = 0 \;\rightarrow\; \Delta x = -H^{-1}\,g. \tag{3.62}$$

The Newton method has a quadratic rate of convergence. In the Levenberg-Marquardt method an additional constraint on the step is imposed $||\Delta x||_2 = 1$. This leads to the Lagrangian function

$$\mathcal{L}(\Delta x, \lambda) = C(x_0) + g^T \Delta x + \frac{1}{2} \Delta x^T H \Delta x + \frac{1}{2} \lambda (\Delta x^T \Delta x - 1). \tag{3.63}$$

The condition for optimality gives

$$\frac{\partial \mathcal{L}}{\partial(\Delta x)} = g + H \Delta x + \lambda \Delta x = 0 \;\rightarrow\; \Delta x = -[H + \lambda I]^{-1} g. \tag{3.64}$$

Hence for $\lambda = 0$ this reduces to the Newton method. For large $\lambda$ it corresponds to a steepest descent method.

A disadvantage of the Newton method is that the second order derivatives in the full Hessian are to be calculated and to be stored. Therefore in quasi-Newton methods one tries to estimate the Hessian based upon first order gradient information only. From (3.61) it follows that

$$H\, d_t = y_t \tag{3.65}$$

with $d_t = x_{t+1} - x_t$ and $y_t = g_{t+1} - g_t$. This means that there is a linear mapping between changes in the gradient and changes in position. Assume now that the cost function $C$ is nonquadratic and that $B_t$ is an estimate for the Hessian. Then we have the update $B_{t+1} = B_t + \Delta B$ and the so-called quasi-Newton condition

$$B_{t+1}\, d_t = y_t \tag{3.66}$$

has to be satisfied. Rank 1 and rank 2 updates of the Hessian are considered then for building up the curvature information. In the rank 1 update $\Delta B = q\, z z^T$, where the scalar $q$ and the vector $z$ follow from the quasi-Newton condition (3.66), one obtains the update

$$B_{t+1} = B_t + \frac{(y_t - B_t d_t)(y_t - B_t d_t)^T}{(y_t - B_t d_t)^T d_t}. \tag{3.67}$$

For the rank 2 update $\Delta B = q_1\, z_1 z_1^T + q_2\, z_2 z_2^T$, where the scalars $q_1$, $q_2$ and the vectors $z_1$, $z_2$ are again determined from the quasi-Newton condition one obtains the well-known BFGS formula (Broyden, Fletcher, Goldfarb, Shanno)

$$B_{t+1} = B_t + \frac{y_t y_t^T}{y_t^T d_t} - \frac{(B_t d_t)(B_t d_t)^T}{(B_t d_t)^T d_t} \tag{3.68}$$

The formulas (3.67) and (3.68) are called direct update formulas, because they update the Hessian. In inverse updating formulas the inverse of the Hessian is updated according to the quasi-Newton condition

$$d_t = R_{t+1}\, y_t \tag{3.69}$$

leading to the well known DFP formula (Davidon, Fletcher, Powell)

$$R_{t+1} = R_t + \frac{d_t d_t^T}{d_t^T y_t} - \frac{R_t y_t y_t^T R_t}{y_t^T R_t y_t}. \tag{3.70}$$

Once the search direction has been found a line search procedure is needed in order to find an optimal step size along the line of the search direction. This is usually done by means of quadratic or cubic line search. Furthermore quasi-Newton methods converge super-linearly.

Another optimization method without updating of matrices is the conjugate gradient algorithm. The scheme is

$$x_{t+1} = x_t + \alpha_t u_t \tag{3.71}$$

with $u_t$ the search direction, which is determined by

$$u_{t+1} = -g_{t+1} + \beta_t u_t. \tag{3.72}$$

Here $\beta_t$ is chosen e.g. as

$$\beta_t = \frac{g_{t+1}^T g_{t+1}}{g_t^T g_t} \tag{3.73}$$

which is the Fletcher-Reeves method or

$$\beta_t = \frac{g_{t+1}^T (g_{t+1} - g_t)}{g_t^T g_t} \tag{3.74}$$

which is the Polak-Ribière method. Initially one starts with $u_0 = -g_0$. Remark that a standard backpropagation method with momentum term corresponds to constant $\alpha$ and $\beta$ values. A typical feature of the conjugate gradient method is that the current search direction is perpendicular to the gradient at the next point

$$u_t^T g_{t+1} = 0 \tag{3.75}$$

and that the search directions $u_t$ and $u_{t+1}$ are conjugated with respect to the Hessian $H$

$$u_t^T H u_{t+1} = 0. \tag{3.76}$$

In case the cost function is purely quadratic the conjugate gradient method converges in $n$ steps for an $n$-dimensional search space. In the non-quadratic case a restart procedure is often applied (Powell (1977)) and accurate line search is needed. Like quasi-Newton methods, conjugate gradient algorithms have a superlinear speed of convergence. In large dimensional search spaces conjugate gradient methods are to be preferred over Newton methods. Conjugate gradient algorithms were applied successfully to neural networks e.g. in Møller (1993), van der Smagt (1994), Charalambous (1992), Johansson *et al.* (1992).

In neural network problems the cost function has often the form of a sum of squared residuals, such as in prediction error algorithms for series-parallel models. The optimization problem is then a nonlinear least squares problem

$$\min_{x \in \mathbb{R}^n} \phi(x) = \frac{1}{2} \sum_{i=1}^{N} r_i(x)^2 \tag{3.77}$$

with residuals

$$r_i(x) = F(u_i; x) - y_i \tag{3.78}$$

where the nonlinear mapping $F(.)$ is e.g. a neural network architecture such as (3.6) and data $\{u_i, y_i\}_{i=1}^{i=N}$ are given. Suppose here $u \in \mathbb{R}^p$, $y \in \mathbb{R}$ and $r(x) = [r_1; r_2; ...; r_N]$. Then the gradient of $\phi$ is

$$\nabla\phi(x) = J(x)^T r(x) \tag{3.79}$$

and the Hessian for $\phi$ is

$$H(x) = J(x)^T J(x) + \sum_{i=1}^{N} r_i H_i(x) \tag{3.80}$$

where $[J_{ij}] = \frac{\partial r_i}{\partial x_j}$ is the Jacobian matrix of $r(x)$ and $H_i(x)$ is the Hessian matrix related to $r_i(x)$. A good approximation for the Hessian is normally $H(x) \simeq J(x)^T J(x)$. In Table 3.2 several search directions are shown for the nonlinear least squares problem for several optimization methods (see Saarinen *et al.* (1993)).

| Algorithm | Search direction |
|---|---|
| Steepest descent | $-J^T r$ |
| Conjugate gradient | $-J^T r + \beta p$ |
| Newton | $-(J^T J + \sum_i r_i H_i)^{-1} J^T r$ |
| Gauss-Newton | $-(J^T J)^{-1} J^T r$ |
| Levenberg-Marquardt | $-(J^T J + \rho_k I)^{-1} J^T r$ |
| Quasi-Newton | $-(J^T J + B_k)^{-1} J^T r$ |

Table 3.2: *Nonlinear least squares: search directions according to several possible optimization methods. In the conjugate gradient method p is the previous search direction; in quasi-Newton $B_k$ satisfies the quasi-Newton condition.*

## 3.5 Aspects of model validation, pruning and regularization

Once a neural network model has been identified from the training data, further model validation and cross validation is needed. Model validity tests for nonlinear models were developed e.g. in Billings & Voon (1986). For prediction error algorithms it is shown that the prediction error $\epsilon_k$ should be uncorrelated with all linear and nonlinear combinations of past inputs and outputs and the following conditions should hold

$$\begin{array}{rcl} \phi_{\epsilon\epsilon}(\tau) & = & E[\epsilon_{k-\tau}\epsilon_k] = \delta_\tau \\ \phi_{u\epsilon}(\tau) & = & E[u_{k-\tau}\epsilon_k] = 0, \ \forall\tau \\ \phi_{u^{2'}\epsilon}(\tau) & = & E[(u_{k-\tau}^2 - \overline{u}_k^2)\epsilon_k] = 0, \ \forall\tau \\ \phi_{u^{2'}\epsilon^2}(\tau) & = & E[(u_{k-\tau}^2 - \overline{u}_k^2)\epsilon_k^2] = 0, \ \forall\tau \\ \phi_{\epsilon(\epsilon u)}(\tau) & = & E[\epsilon_k\epsilon_{k-1-\tau}u_{k-1-\tau}] = 0, \ \forall\tau \geq 0 \end{array} \tag{3.81}$$

where $u_k^{2'} = u_k^2 - \overline{u_k^2}$ and $\overline{u_k^2}$ denotes the mean of $u_k^2$ in time. In practice one works with normalized correlations

$$\hat{\phi}_{\psi\varphi}(\tau) = \frac{\sum_{t=1}^{N-\tau} \psi_k \varphi_{k+\tau}}{[\sum_{k=1}^{N} \psi_k^2 \sum_{k=1}^{N} \varphi_k^2]^{1/2}} \tag{3.82}$$

for two sequences $\psi_k, \varphi_k$, such that $-1 \leq \hat{\phi}_{\psi\varphi}(\tau) \leq 1$. If the correlation functions are within 95 % confidence intervals ($\pm 1.96/\sqrt{N}$), then the model can be regarded as adequate. If the conditions (3.81) hold, the estimate will be unbiased. The conditions are applicable to the measurement noise as well as the process noise case and can be applied both to the training set and the test set. Validation means checking on the training data and cross validation on the test data ('fresh' data) (see Ljung (1987), Chen *et al.* (1990a)). Other statistical tests such as the chi-squared test (see Chen *et al.* (1990a)) can also be employed.

As suggested by Akaike's information criterion one should try to obtain a best fit with a minimal number of parameters (see Ljung (1987)). Hence on the one hand one needs 'enough' hidden neurons in order to model the possibly complex underlying nonlinear dynamics but on the other hand not too much hidden neurons in order to obtain a good generalization of the model. When too many parameters are taken, a typical behaviour of the error on the training set and the test set is shown in Figure 3.3 during optimization. At the beginning the error on the training set and test set both decrease. At a certain moment the error on the test set starts increasing while the error on the training set still decreases. From that time overfitting occurs. A procedure which is often adopted in neural networks is then to stop learning once the test set error starts increasing (Hammerstrom (1993)). In the work of Sjöberg & Ljung (1992)(1995) this procedure is theoretically justified for prediction error algorithm related to the NARX model (3.6). Assuming there is a true system $\theta_0$ that satisfies

$$y_k = f(x_{k-1}; \theta_0) + e_k \tag{3.83}$$

and letting $\lambda_0 = Ee_k^2$, then

$$\overline{V}_N = E\overline{V}(\hat{\theta}_N) = \lambda_0(1 + \frac{d}{2N}) \tag{3.84}$$

where $\overline{V}(\theta) = E(y_k - f(x_{k-1};\theta))$, $\hat{\theta}_N = \arg\min_\theta V_N(\theta)$, $d = \dim\theta$ and $N$ is the number of data points. From the expression (3.84) it becomes clear that too many parameters have a bad influence. If one considers instead the regularized problem

$$W_N(\theta) = V_N(\theta) + \frac{1}{2}\delta\,|\theta - \theta_0|^2 \tag{3.85}$$

with regularization parameter $\delta$, then it can be shown that

$$\overline{V}_N^\delta = E\overline{V}(\hat{\theta}_N^\delta) = \lambda_0(1 + \frac{\tilde{d}}{2N}) \tag{3.86}$$

where

$$\tilde{d} = \sum_{i=1}^{d} \frac{\lambda_i^2}{(\lambda_i^2 + \delta)^2} \tag{3.87}$$

where $\lambda_i$ are the eigenvalues of the matrix $Q = E\psi_k\psi_k^T$ with $\psi_k = \frac{\partial f}{\partial \theta}|_{\theta=\theta_0}$ and $\hat{\theta}_N^\delta$ is the minimizer of $W_N(\theta)$. Hence the superfluous parameters no longer have a bad influence on the model performance. Pragmatically one can think of $\tilde{d}$ as the efficient number of parameters, i.e. those that contribute to the Hessian with eigenvalues larger than $\delta$. In fact because the value of $\theta_0$ is unknown one has to take some guess $\theta^*$ for $\theta_0$ in (3.85), which leads to an additional correction to (3.86). It is shown then that

$$\hat{\theta}_N^\delta = (I - M_\delta)\hat{\theta}_N + M_\delta\theta^* \tag{3.88}$$

where $M_\delta = \delta(\delta I + Q)^{-1}$. Based on the expression (3.88) it is shown in Sjöberg & Ljung (1992) that doing an 'unfinished' search in the minimization of $V_N(\theta)$, has a similar effect as regularizing towards $\theta^*$.

In pruning methods (see e.g. Reed (1993) for a survey) the approach is to train a network that is larger than required and remove the unnecessary parameters, after or during training. One makes a distinction between sensitivity calculations and penalty term methods. In sensitivity methods the training is as usual but after training one looks at the sensitivity of each interconnection weight with respect to the cost function, by taking the value of the weight after training and upon its removal. The weights with low sensitivity are pruned then. In penalty term methods the original cost function is modified (like this is done in the regularized problem). Examples are e.g. a weight decay method, where the cost function is adapted into

$$E = \sum_p E_p + \lambda \sum_{ij} |w_{ij}| \tag{3.89}$$

with $\lambda$ a positive constant, and the method described in Weigend *et al.* (1991)

$$E = \sum_p E_p + \lambda \sum_{ij} \frac{(w_{ij}/w_0)^2}{1 + (w_{ij}/w_0)^2}. \tag{3.90}$$

For $|w_{ij}| \gg w_0$, the cost of a weight approaches $\lambda$ and for $|w_{ij}| \ll w_0$ the cost is nearly zero. This method requires some tuning of the parameters $\lambda$ and $w_0$.

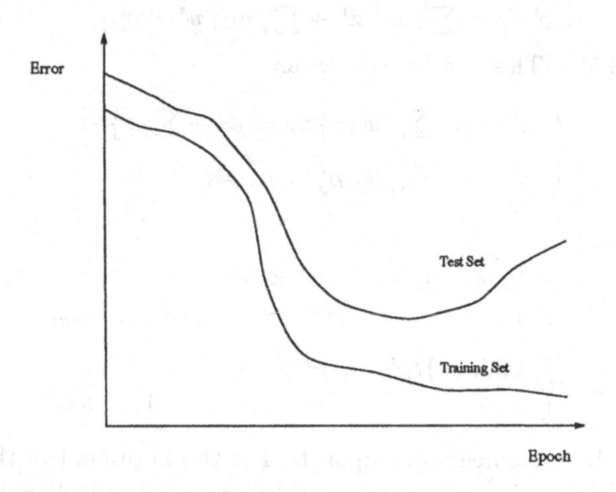


Figure 3.3: *When the proposed model has too many parameters, overfitting may occur. The error on the training set keeps on decreasing during training, while the error on the test set (fresh data) starts increasing at a certain moment. Training is usually stopped when the minimal test set error is achieved.*

## 3.6 Neural network models as uncertain linear systems

In this Section we will explain how neural network models can be interpreted as uncertain linear models. The ideas described here are introduced in Suykens *et al.* (1994b). An example is given in Section 3.7.2.

### 3.6.1 Convex polytope

Let us consider again the neural state space model (3.14)

$$\begin{cases} \hat{x}_{k+1} & = & W_{AB} \tanh(V_A \hat{x}_k + V_B u_k + \beta_{AB}) + K\epsilon_k \ ; \ \hat{x}_0 = x_0 \\ \hat{y}_k & = & W_{CD} \tanh(V_C \hat{x}_k + V_D u_k + \beta_{CD}) \end{cases}$$

or in elementwise notation

$$\begin{cases} \hat{x}^i & := & \sum_j w_{AB}{}^i_j \tanh(\phi^j) + \sum_j \kappa^i_j \epsilon^j \\ \hat{y}^i & = & \sum_j w_{CD}{}^i_j \tanh(\rho^j) \end{cases} \tag{3.91}$$

with

$$\begin{array}{rcl} \phi^l & = & \sum_r v_{A}{}^l_r \hat{x}^r + \sum_s v_{B}{}^l_s u^s + \beta^l_{AB} \\ \rho^l & = & \sum_r v_{C}{}^l_r \hat{x}^r + \sum_s v_{D}{}^l_s u^s + \beta^l_{CD} \end{array} \tag{3.92}$$

according to (3.53). This can be written as

$$\begin{cases} \hat{x}^i & := & \sum_j w_{AB}{}^i_j \gamma_{AB}{}^j_j \phi^j + \sum_j \kappa^i_j \epsilon^j \\ \hat{y}^i & = & \sum_j w_{CD}{}^i_j \gamma_{CD}{}^j_j \rho^j , \end{cases} \tag{3.93}$$

with

$$\begin{array}{l} \gamma_{AB}{}^j_j = \begin{cases} \tanh(\phi^j)/\phi^j & ,(\phi^j \neq 0) \\ 1 & ,(\phi^j = 0) \ ; \ j = 1, ..., n_{hx} \end{cases} \\ \gamma_{CD}{}^j_j = \begin{cases} \tanh(\rho^j)/\rho^j & ,(\rho^j \neq 0) \\ 1 & ,(\rho^j = 0) \ ; \ j = 1, ..., n_{hy}. \end{cases} \end{array} \tag{3.94}$$

The fact that the $\gamma$ elements are equal to 1 if the argument of the activation function becomes zero is easily seen by applying de l' Hospital's rule or by using the Taylor expansion for tanh(.). Essential is that the activation function is a static nonlinearity that belongs to the sector [0, 1]. The $\gamma$ elements have the property that they belong to bounded intervals: $\gamma_{AB}{}^j_j \in [0, 1]$, $\gamma_{CD}{}^j_j \in [0, 1]$ for all $j$. Turning back to the matrix-vector notation, (3.14) can be written as

$$\begin{cases} \hat{x}_{k+1} & = & A(\hat{x}_k, u_k)\hat{x}_k + B(\hat{x}_k, u_k)u_k + v(\hat{x}_k, u_k) + K\epsilon_k \\ \hat{y}_k & = & C(\hat{x}_k, u_k)\hat{x}_k + D(\hat{x}_k, u_k)u_k + w(\hat{x}_k, u_k) \end{cases} \tag{3.95}$$

where

$$\begin{array}{ll} A(\hat{x}_k, u_k) = W_{AB}\Gamma_{AB}(\hat{x}_k, u_k)V_A & B(\hat{x}_k, u_k) = W_{AB}\Gamma_{AB}(\hat{x}_k, u_k)V_B \\ C(\hat{x}_k, u_k) = W_{CD}\Gamma_{CD}(\hat{x}_k, u_k)V_C & D(\hat{x}_k, u_k) = W_{CD}\Gamma_{CD}(\hat{x}_k, u_k)V_D \\ v(\hat{x}_k, u_k) = W_{AB}\Gamma_{AB}(\hat{x}_k, u_k)\beta_{AB} & w(\hat{x}_k, u_k) = W_{CD}\Gamma_{CD}(\hat{x}_k, u_k)\beta_{CD} \end{array} \tag{3.96}$$

and diagonal matrices

$$\begin{aligned} \Gamma_{AB} &= \text{diag}\{\gamma_{AB}{}_1^1, ..., \gamma_{AB}{}_{n_{hx}}^{n_{hx}}\} \\ \Gamma_{CD} &= \text{diag}\{\gamma_{CD}{}_1^1, ..., \gamma_{CD}{}_{n_{hy}}^{n_{hy}}\}. \end{aligned} \tag{3.97}$$

Note that for zero bias terms in the neural state space model the terms $v$ and $w$ become zero. The matrices in (3.95) may be written as

$$\begin{aligned} A &= \sum_{i=1}^{n_{hx}} \gamma_{AB_i} A_i & A_i &= W_{AB}(:,i)\, V_A(i,:) \\ B &= \sum_{i=1}^{n_{hx}} \gamma_{AB_i} B_i & B_i &= W_{AB}(:,i)\, V_B(i,:) \\ v &= \sum_{i=1}^{n_{hx}} \gamma_{AB_i} v_i & v_i &= W_{AB}(:,i)\, \beta_{AB}(i) \\ C &= \sum_{i=1}^{n_{hy}} \gamma_{CD_i} C_i & C_i &= W_{CD}(:,i)\, V_C(i,:) \\ D &= \sum_{i=1}^{n_{hy}} \gamma_{CD_i} D_i & D_i &= W_{CD}(:,i)\, V_D(i,:) \\ w &= \sum_{i=1}^{n_{hy}} \gamma_{CD_i} w_i & w_i &= W_{CD}(:,i)\, \beta_{CD}(i). \end{aligned} \tag{3.98}$$

Hence the matrices belong to the convex hull

$$[A\ B\ v] \in \text{Co}\{[A_1\ B_1\ v_1], ..., [A_r\ B_r\ v_r]\} \tag{3.99}$$

which is a convex polytope with $2^r$ vertices (corners) and $r = n_{hx}$ and

$$[C\ D\ w] \in \text{Co}\{[C_1\ D_1\ w_1], ..., [C_s\ D_s\ w_s]\} \tag{3.100}$$

a convex polytope with $2^s$ vertices and $s = n_{hy}$.

A similar interpretation can be given to input/output models. Let us take the NARX model (3.6)

$$y_k = w^T \tanh(V x_{k-1}) + e_k.$$

This can be written as an uncertain ARX model as

$$y_k = [a(x_{k-1})]^T x_{k-1} + e_k \tag{3.101}$$

with uncertain parameter vector

$$[a(x_{k-1})]^T = w^T \Gamma(x_{k-1}) V \tag{3.102}$$

and $\Gamma = \text{diag}\{\gamma_i\}$ with a number of diagonal elements $\gamma_i = \frac{\tanh(v_i^T x_{k-1})}{v_i^T x_{k-1}}$ equal to the number of hidden neurons in the neural network.

The interpretation of a neural network model as an uncertain linear system has some nice consequences

1. *Choice of specific parametrizations:*
Neural network models are described by a lot of parameters and are mostly used as parametrizations in black box models. The objection is then often heard that one looses much physical insight into the problem. The interpretation as an uncertain linear system might overcome this problem up to a certain extent. It is possible to interpret the system with a lower number of parameters equal to that of a linear system and to impose more structure in the parametrization depending on the physical insight and a priori information on the system. If one knows e.g. from physical considerations that the system has constant $B$,$C$,$D$ matrices and a state dependent $A$ matrix in (3.95), it will be better to prefer a more specific parametrization like e.g.

$$\left\{ \begin{array}{rcl} \hat{x}_{k+1} & = & W_A \tanh(V_A \hat{x}_k) + B u_k + K \epsilon_k \ ; \ \hat{x}_0 = x_0 \\ \hat{y}_k & = & C \hat{x}_k + D u_k \end{array} \right. \quad (3.103)$$

instead of the full parametrization in (3.14), which also reduces the number of parameters to be estimated, leading to better generalization according to (3.84).

2. *Estimating 'hardness' of distortion:*
The $\gamma$ elements give a qualitative indication of the 'hardness' of distortion of the underlying nonlinearity of the system. For given input signals $u_k$ the $\gamma$ elements can be calculated through time. If they are close to 1 then the model is close to linear. If they come close to 0 then the model is highly nonlinear, because the activation functions of the neural networks are going into saturation.

3. *Linear models as starting point for optimization:*
In case $\Gamma_{AB} = I$ and $\Gamma_{CD} = I$ we obtain a purely linear system with system matrices

$$\left[ \begin{array}{cc} A & B \\ C & D \end{array} \right] = \left[ \begin{array}{cc} W_{AB} V_A & W_{AB} V_B \\ W_{CD} V_C & W_{CD} V_D \end{array} \right]. \quad (3.104)$$

This observation yields a procedure for taking linear state space models as possible initial estimates for the neural state space model in the optimization problem. Suppose a linear state space model in innovations form is available (Ljung (1987))

$$\left\{ \begin{array}{rcl} \hat{x}_{k+1} & = & A \hat{x}_k + B u_k + K \epsilon_k \\ \hat{y}_k & = & C \hat{x}_k + D u_k \end{array} \right. \quad (3.105)$$

with $E\{\epsilon_k \epsilon_s^T\} = \Lambda \delta_{ks}$ ($\Lambda$ diagonal). Then by taking

$$W_{AB} := \frac{1}{\alpha_1}[I_n \; R_1], \quad [V_A \; V_B] := \alpha_1 \begin{bmatrix} A & B \\ 0 & 0 \end{bmatrix}, \; \beta_{AB} := 0; \; (n_{hx} \geq n)$$

$$W_{CD} := \frac{1}{\alpha_2}[I_l \; R_2], \quad [V_C \; V_D] := \alpha_2 \begin{bmatrix} C & D \\ 0 & 0 \end{bmatrix}, \; \beta_{CD} := 0; \; (n_{hy} \geq l) \tag{3.106}$$

and letting $\alpha_1, \alpha_2 \to 0$, one enforces the neural state space model to behave linearly. $R_1, R_2$ are arbitrary matrices of appropriate dimension. A similar procedure can be applied for I/O models (Suykens & De Moor (1993b)). Especially subspace algorithms are useful for generating these starting points, because they are one-shot non-iterative methods for identifying linear state space models (Moonen *et al.* (1989), De Moor *et al.* (1991), Van Overschee & De Moor (1994)).

4. *Use of neural network models within robust control theory:*
   In robust control theory, methods are available for designing linear controllers for uncertain linear models. In the next subsection the uncertain linear system will be brought into a form which fits immediately in the framework of $\mu$ control theory.

### 3.6.2 LFT representation

One way of representing uncertainties is by means of Linear Fractional Transformations (LFTs) which consists of defining a nominal linear model and modelling uncertainty through a feedback perturbation on the nominal model, as is shown in Figure 3.4. (see e.g. Maciejowski (1989), Boyd & Barratt (1991)). The LFT is obtained as follows. Depending on the input and state vector sequences the elements ${\gamma_{AB}}_j^j$ and ${\gamma_{CD}}_j^j$ belong to intervals

$$\begin{aligned} {\gamma_{AB}}_j^j &\in [{\gamma_{AB}^-}_j^j, {\gamma_{AB}^+}_j^j] &\subset [0,1] \\ {\gamma_{CD}}_j^j &\in [{\gamma_{CD}^-}_j^j, {\gamma_{CD}^+}_j^j] &\subset [0,1]. \end{aligned} \tag{3.107}$$

The nominal $\gamma$ values are then defined as the midpoint of these intervals. The choice of these intervals is a matter of degree of conservativeness that one wants to take into account and will depend on the input signals for which the predictor is simulated. The most conservative LFT will be obtained by setting all nominal $\gamma$ values equal to 0.5 as the midpoint of the intervals [0,1], ensuring independence of the input and state vector sequence. According to Steinbuch *et al.* (1992) the following definitions are made

$$\begin{aligned} \gamma^{(0)}{}_{ABj}^{\,j} &= (\gamma^{-}{}_{ABj}^{\,j} + \gamma^{+}{}_{ABj}^{\,j})/2 & s_{AB}{}_j^j &= (\gamma^{+}{}_{ABj}^{\,j} - \gamma^{-}{}_{ABj}^{\,j})/2 \\ \gamma^{(0)}{}_{CDj}^{\,j} &= (\gamma^{-}{}_{CDj}^{\,j} + \gamma^{+}{}_{CDj}^{\,j})/2 & s_{CD}{}_j^j &= (\gamma^{+}{}_{CDj}^{\,j} - \gamma^{-}{}_{CDj}^{\,j})/2 \end{aligned} \tag{3.108}$$

and

$$\begin{aligned} \Gamma^{(0)}_{AB} &= \mathrm{diag}\{\gamma^{(0)}{}_{\mathrm{AB}1}^{\,1}, ..., \gamma^{(0)}{}_{\mathrm{ABn_{hx}}}^{\,\mathrm{n_{hx}}}\} & S_{AB} &= \mathrm{diag}\{\mathrm{s_{AB}}{}_1^1, ..., \mathrm{s_{AB}}{}_{\mathrm{n_{hx}}}^{\mathrm{n_{hx}}}\} \\ \Gamma^{(0)}_{CD} &= \mathrm{diag}\{\gamma_{\mathrm{CD}1}^{\,1}, ..., \gamma^{(0)}{}_{\mathrm{CDn_{hy}}}^{\,\mathrm{n_{hy}}}\} & S_{CD} &= \mathrm{diag}\{\mathrm{s_{CD}}{}_1^1, ..., \mathrm{s_{CD}}{}_{\mathrm{n_{hy}}}^{\mathrm{n_{hy}}}\}. \end{aligned} \tag{3.109}$$

We have then

$$\begin{aligned} \Gamma_{AB} &= \Gamma^{(0)}_{AB} + S_{AB}\,\Delta_{AB} & \Delta_{AB} &= \mathrm{diag}\{\delta_{\mathrm{AB}}{}_1^1, ..., \delta_{\mathrm{AB}}{}_{\mathrm{n_{hx}}}^{\mathrm{n_{hx}}}\} \\ \Gamma_{CD} &= \Gamma^{(0)}_{CD} + S_{CD}\,\Delta_{CD} & \Delta_{CD} &= \mathrm{diag}\{\delta_{\mathrm{CD}}{}_1^1, ..., \delta_{\mathrm{CD}}{}_{\mathrm{n_{hy}}}^{\mathrm{n_{hy}}}\} \end{aligned} \tag{3.110}$$

with $\delta_{AB}{}_j^j \in [-1,1]$ and $\delta_{CD}{}_j^j \in [-1,1]$, such that $\|\Delta_{AB}\| \leq 1$ and $\|\Delta_{CD}\| \leq 1$. The matrices (3.96) can then be written as

$$\begin{aligned} A(\hat{x}_k, u_k) &= A^{(0)} + A_\delta(\delta_{AB}) & B(\hat{x}_k, u_k) &= B^{(0)} + B_\delta(\delta_{AB}) \\ C(\hat{x}_k, u_k) &= C^{(0)} + C_\delta(\delta_{CD}) & D(\hat{x}_k, u_k) &= D^{(0)} + D_\delta(\delta_{CD}) \\ v(\hat{x}_k, u_k) &= v^{(0)} + v_\delta(\delta_{AB}) & w(\hat{x}_k, u_k) &= w^{(0)} + w_\delta(\delta_{CD}) \end{aligned} \tag{3.111}$$

with

$$\begin{aligned} A^{(0)} &= W_{AB}\Gamma^{(0)}_{AB}V_A & A_\delta(\delta_{AB}) &= W_{AB}S_{AB}\Delta_{AB}V_A \\ B^{(0)} &= W_{AB}\Gamma^{(0)}_{AB}V_B & B_\delta(\delta_{AB}) &= W_{AB}S_{AB}\Delta_{AB}V_B \\ C^{(0)} &= W_{CD}\Gamma^{(0)}_{CD}V_C & C_\delta(\delta_{CD}) &= W_{CD}S_{CD}\Delta_{CD}V_C \\ D^{(0)} &= W_{CD}\Gamma^{(0)}_{CD}V_D & D_\delta(\delta_{CD}) &= W_{CD}S_{CD}\Delta_{CD}V_D \\ v^{(0)} &= W_{AB}\Gamma^{(0)}_{AB}\beta_{AB} & v_\delta(\delta_{AB}) &= W_{AB}S_{AB}\Delta_{AB}\beta_{AB} \\ w^{(0)} &= W_{CD}\Gamma^{(0)}_{CD}\beta_{CD} & w_\delta(\delta_{CD}) &= W_{CD}S_{CD}\Delta_{CD}\beta_{CD}. \end{aligned} \tag{3.112}$$

An LFT is obtained then for the neural state space model

$$\hat{y} = \mathcal{F}_u(G, \Delta)\begin{bmatrix} u \\ \epsilon \\ 1 \end{bmatrix} \tag{3.113}$$

with state space representation

$$\left\{ \begin{array}{lcl} G: \begin{bmatrix} \hat{x}_{k+1} \\ \hline \hat{y}_k \\ q_k \end{bmatrix} & = & \left[ \begin{array}{c|ccccc} A^{(0)} & B^{(0)} & K & v^{(0)} & [W_{AB}S_{AB} \ 0] \\ \hline C^{(0)} & D^{(0)} & 0 & w^{(0)} & [0 \ W_{CD}S_{CD}] \\ \begin{bmatrix} V_A \\ V_C \end{bmatrix} & \begin{bmatrix} V_B \\ V_D \end{bmatrix} & 0 & \begin{bmatrix} \beta_{AB} \\ \beta_{CD} \end{bmatrix} & 0 \end{array} \right] \cdot \begin{bmatrix} \hat{x}_k \\ \hline u_k \\ \epsilon_k \\ 1 \\ p_k \end{bmatrix} \\ \\ \Delta: \quad p_k & = & \Delta . q_k \quad , \Delta = \text{diag}\{\Delta_{AB}, \Delta_{CD}\} \quad , ||\Delta|| \leq 1. \end{array} \right. \tag{3.114}$$

At the component level the uncertainty is caused by the $\gamma$ elements and their corresponding $\delta$ elements which depend nonlinearly on the input and state vector. The interconnection matrices of the neural state space model are assumed to be exact here. At the system level this uncertainty becomes *structured* because the matrix $\Delta$ is diagonal (see e.g. Maciejowski (1989), Packard & Doyle (1993)). The uncertainty is *real* and the dimension of $\Delta$ does only depend upon the number of hidden neurons of the neural networks. In Chapter 5 more aspects on robust control theory and control design will be given.

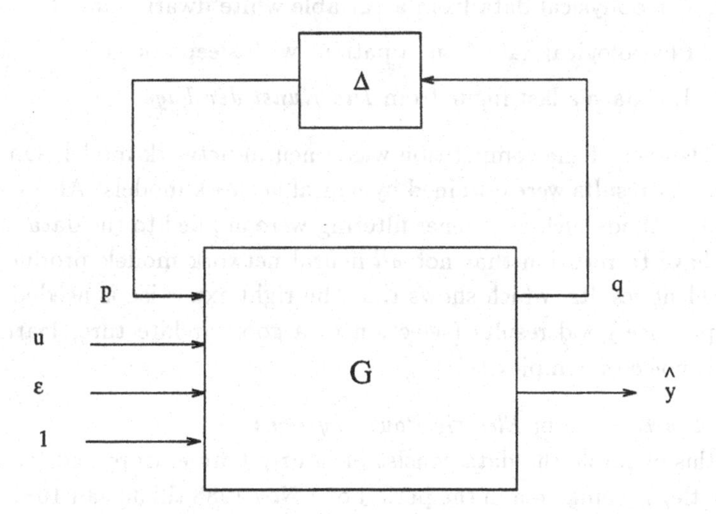


Figure 3.4: ***Linear Fractional Transformation (LFT) representation of a neural state space model. After defining the nominal model G, bounded uncertainty is pulled out in the Δ block, represented in feedback form. The elements of Δ are real and nonlinear in nature. Essentially this representation as an LFT is possible because the activation function*** tanh(.) ***in the neural state space model is a static nonlinearity lying in the sector*** [0, 1].

# 3.7 Examples

## 3.7.1 Some challenging examples from the literature

I/O models such as NARX and NARMAX have already proven to be successful in many applications. Some convincing examples from the literature are

- *The Santa Fe time series prediction contest:*
  From august 1991 onward, a set of time series was made generally available at the Santa Fe institute (Weigend & Gershenfeld (1993)). Participants in the competition were asked to submit forecasts of continuation of the data sets (that were withheld). The following data sets were selected

  1. A clean physics laboratory experiment with a $NH_3$ laser (low-dimensional deterministic chaos)
  2. Computer generated series (high-dimensional deterministic chaos)
  3. Currency exchange rate data (Swiss Franc - US Dollar)
  4. Astrophysical data from a variable white dwarf star
  5. Physiological data from a patient with sleep apnea
  6. J.S. Bach's last fugue from *Die Kunst der Fuge*

  The winner of the competition was a neural network model. On all data sets best results were obtained by neural network models. Also more classical methods such as Wiener filtering were applied to the data. However we have to mention that not *all* neural network models produced these excellent results, which shows that the right expertise is needed in order to produce good results (selection of a good architecture, learning rule and choice of complexity).

- *Load forecasting in Electric Power systems:*
  In this example the data consist of hourly temperature and load for the Seattle/Tacoma area in the period of 1 Nov 1988 till 30 Jan 1989 (Park *et al.* (1991)). Peak load, total load and hourly load were predicted based on the temperature by means of neural network models. Improvements up to 2 % were obtained with respect to the currently used techniques.

- *Modelling of the space shuttle main engine:*
  Although a complete nonlinear dynamic model is already available for simulation of the space shuttle main engine, its complexity is too high for use in real time applications. Because of their massively parallel structure, neural nets are particularly useful for emulation and fast simulation

of the engine's dynamics (Saravanan *et al.* (1993)). The system is multivariable and consists of 5 inputs, which is the rotary motion of five valves and 4 outputs, which are the chamber pressure, the mixture ratio and the speed of high pressure fuel and oxidizer turbine. The identification with neural nets was successful.

- *F-16 aircraft dynamic modeling:*
  The behaviour of the F-16 aircraft is characterized by significant nonlinearities due to kinematic and inertial couplings and aerodynamic nonlinearities. Both low and high angle of attack dynamics are taken into account in order to have full effect of nonlinearities. Neural networks were able here to emulate the dynamical behaviour in highly nonlinear operation mode (Youssef (1993)).

We will focus now on the use of neural state space models in the following examples.

### 3.7.2 Simulated nonlinear system with hysteresis

The system to be identified here is an interconnected system, consisting of two dynamical subsystems and two static nonlinearities: a hysteresis curve $f_1(.)$ and a hyperbolic tangent function $f_2(.) = \tanh(.)$ (see Figure 3.5). The linear systems $L$ and $M$ are both SISO of order 2 and 1 with state vectors $x_k$ and $z_k$ respectively. The interconnected system with input $u_k$, output $y_k$ and state vector $[x_k; z_k]$ has the following form

$$\left\{ \begin{array}{lll} x_{k+1} & = & A_L x_k + b_L u_k + \begin{bmatrix} 1 \\ 0 \end{bmatrix} v_k \\ z_{k+1} & = & a_M z_k + b_M f_1(c_L^T x_k) \\ y_k & = & f_2(c_M z_k + d_M f_1(c_L^T x_k)) + w_k \end{array} \right. \quad (3.115)$$

with $v_k$, $w_k$ zero mean white Gaussian noise processes. The I/O data were generated by a random input signal, uniformly distributed in the interval [-1,1]. Process noise $v_k$ and measurement noise $w_k$ have both standard deviation 0.01. The system matrices for $L$ and $M$ are $a_M = 0.7, b_M = c_M = d_M = 1$

$$A_L = \begin{bmatrix} 0.1 & -0.2 \\ 1 & 0.3 \end{bmatrix}, b_L = \begin{bmatrix} 0 \\ 1 \end{bmatrix}, c_L = \begin{bmatrix} 1 \\ 0 \end{bmatrix}, d_L = 0.$$

The nonlinearity $f_1(.)$ is shown in Figure 3.5 and defined by the following table

| | $x_2 > 0$ | $x_2 \leq 0$ |
|---|---|---|
| $-c-d \leq x_1 \leq -c+d$ | $f_1 = -c$ | - |
| $-c+d \leq x_1 \leq c+d$ | $f_1 = x_1 - d$ | - |
| $c-d \leq x_1 \leq c+d$ | - | $f_1 = c$ |
| $-c-d \leq x_1 \leq c-d$ | - | $f_1 = x_1 + d$ |
| $-c-d \leq x_1$ | $f_1 = -c$ | $f_1 = -c$ |
| $x_1 \leq c+d$ | $f_1 = c$ | $f_1 = c$ |

which means that the right or left part of the curve is selected depending on the sign of $x_2$. In total 2000 data points were generated by (3.115) with $c = 1$, $d = 0.2$ in $f_1$ and zero initial state. This data set is split into two parts: a training set containing the first 1000 data points ($N_{fit} = 1000$) and a test set consisting of the following 1000 data points ($N_{gen} = 1000$) which are fresh data to test the obtained models. The corresponding fitting error $V_{N_{fit}}$ and the generalization error $V_{N_{gen}}$ are defined on these sets. As predictor a neural state space model (3.14) was taken with $n = 3$, $n_{hx} = n_{hy} = 7$. The number of hidden neurons is chosen on a trial and error basis. In order to minimize the cost function (3.47) a quasi-Newton method with BFGS updating of the Hessian and a mixed quadratic and cubic line search was used (function fminu of Matlab's optimization toolbox). Simulation of the neural state space model and its corresponding sensitivity model, needed to generate the gradient of the cost function, were both written in C code, making use of Matlab's *cmex* facility. The best local minimum after taking 100 different starting points (according to a random Gaussian distribution with standard deviation 0.5) was $V_{N_{fit}} = 6.3848\,e-04$. This model has also a minimal generalization error equal to $V_{N_{gen}} = 1.5803\,e-03$. Simulation results for this model are shown in Figure 3.6. The time needed to do these simulations on a DEC 5000/240 workstation is in the order of magnitude of one day. Model validation tests were done according to Billings *et al.* (1992) and are presented in Figure 3.7 for the training data.

For the interpretation of the system as an uncertain linear system, the intervals $[\gamma^{-\,j}_{AB_j}, \gamma^{+\,j}_{AB_j}]$, $[\gamma^{-\,j}_{CD_j}, \gamma^{+\,j}_{CD_j}]$, were calculated based on the training data and the optimal model with as result $\gamma^{+\,j}_{AB_j} = 1$, $\gamma^{+\,j}_{CD_j} = 1$ and

$$\gamma^{-}_{AB} = [0.0562\ \ 0.8942\ \ 0.7313\ \ 0.7659\ \ 0.8904\ \ 0.8306\ \ 0.3841]^T$$

$$\gamma^{-}_{CD} = [0.1785\ \ 0.6271\ \ 0.5012\ \ 0.6022\ \ 0.0244\ \ 0.1877\ \ 0.2284]^T$$

which indicates that the nonlinearity of the underlying system is rather 'hard'. The variation on the elements of the matrix $A(\hat{x}_k, u_k)$ for (3.96) is shown in Figure 3.8. Similar plots can be made for $B(\hat{x}_k, u_k)$, $C(\hat{x}_k, u_k)$, $D(\hat{x}_k, u_k)$.

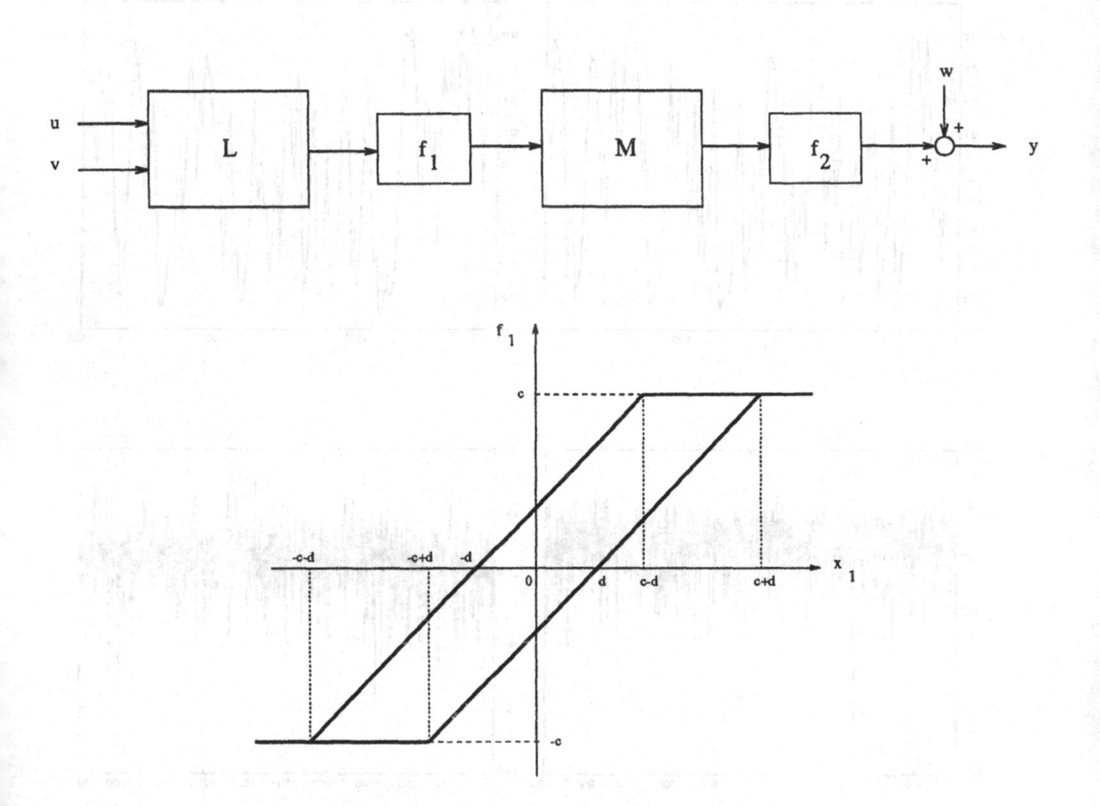


Figure 3.5: *(Top) Nonlinear interconnected system consisting of two linear dynamic systems $L$ and $M$, respectively of order 2 and 1, and two static nonlinearities: a hysteresis curve $f_1(.)$ and $f_2(.) = \tanh(.)$. The system is corrupted with process noise $v$ and measurement noise $w$. (Bottom) Hysteresis curve $f_1(x_1)$. Depending on the sign of $x_2$ the right or left part of the curve is selected.*

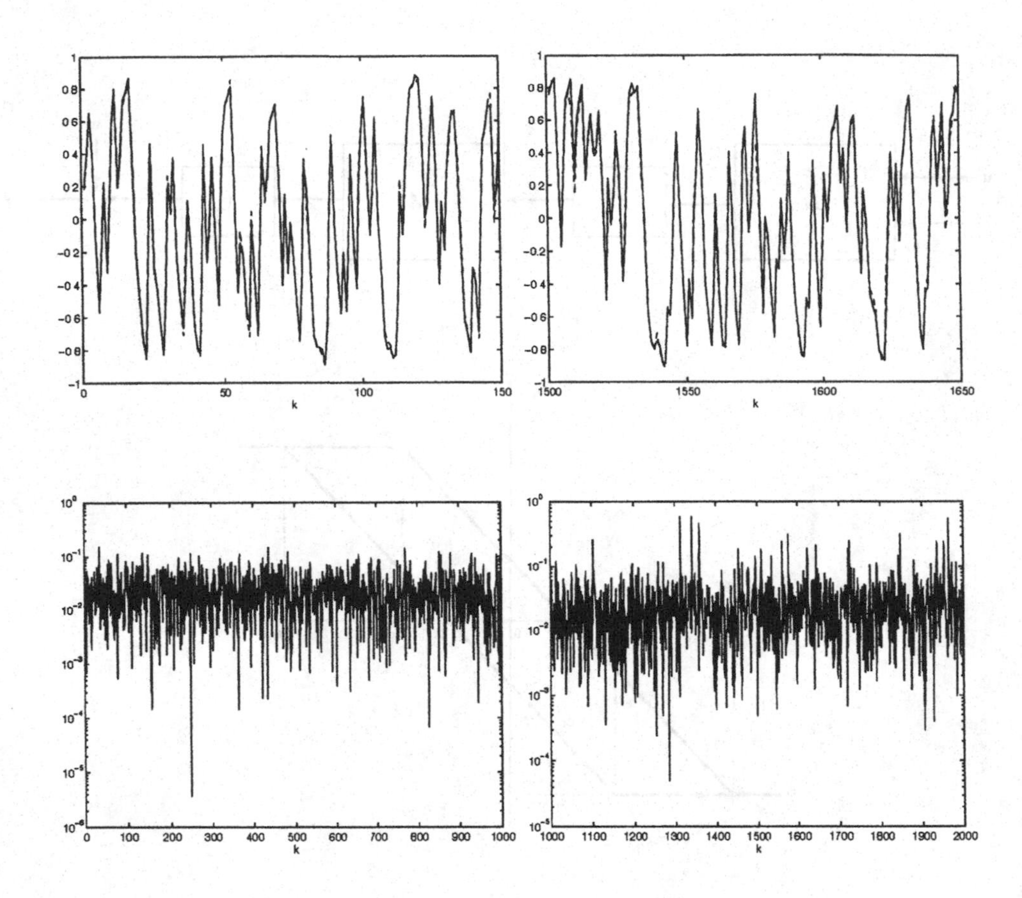

Figure 3.6: *Nonlinear system identification using neural state space models: (Top) output of the plant (full line) and output of the neural state space model (dashed line) through time, shown for part of the training data (Left) and part of the test data (Right); (Bottom) absolute value of the prediction error through time, for the training data (Left) and the test data (Right).*

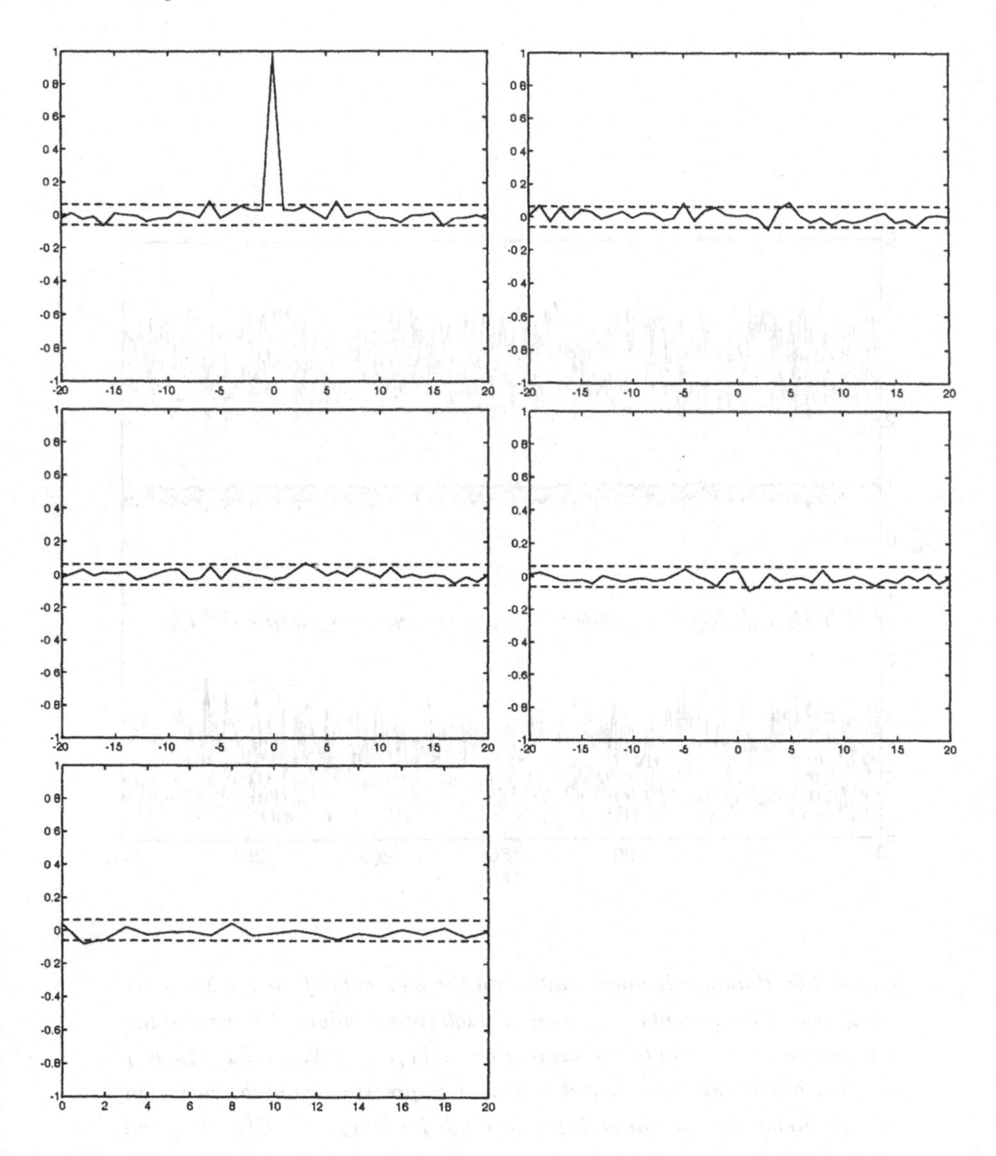


Figure 3.7: *Model validation: normalized correlation tests with* 95% *confidence intervals for the model with minimal fitting and generalization error, evaluated on the training set: a/* $\hat{\phi}_{\epsilon\epsilon}(\tau)$, *b/* $\hat{\phi}_{u\epsilon}(\tau)$, *c/* $\hat{\phi}_{u^{2'}\epsilon}(\tau)$, *d/* $\hat{\phi}_{u^{2'}\epsilon^2}(\tau)$, *e/* $\hat{\phi}_{\epsilon(\epsilon u)}(\tau)$.

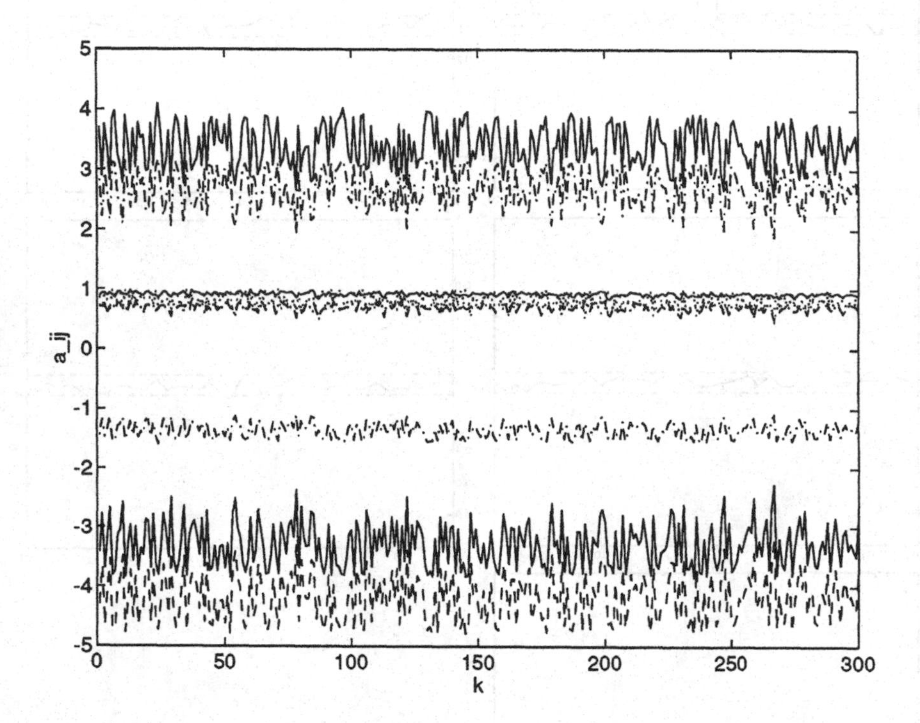


Figure 3.8: *Parametric uncertainties on the elements of the* $3 \times 3$ *matrix* $A(\hat{x}_k, u_k)$. *The elements are shown through time, evaluated on part of the training set, according to the expression* $A(\hat{x}_k, u_k) = W_{AB}\ \Gamma_{AB}(\hat{x}_k, u_k)\ V_A$. *The aim of this figure is just to give an impression about the variation of these elements. Similar plots can be made for* $B(\hat{x}_k, u_k)$, $C(\hat{x}_k, u_k)$ *and* $D(\hat{x}_k, u_k)$.

### 3.7.3 Identification of a glass furnace

The system considered here is a glass feeder of Philips Eindhoven, consisting of three inputs (two heaters and one cooler), switched by pseudo-random binary noise and six outputs, which are temperatures in a cross-section of the oven. Linear system identification of this system was already extensively studied in the past (see e.g. De Moor (1988)). The aim here is to show how results from linear system identification can be used as starting points for neural state space models, in order to produce more accurate nonlinear models. In this example, only the transfer from the three inputs to the first output is modelled. A training set of 500 data points and a test set of the next 500 data points was defined. First linear deterministic identification for the training data was done using the subspace algorithm described in Moonen *et al.* (1989) (see Van Overschee & De Moor (1994) for more advanced algorithms). A third order model was selected. The error on the training set was $4.8522\,e-02$ and on the test set $1.2052\,e-01$. Then the neural state space model (3.14) was initialized by means of the linear model with zero Kalman gain, zero bias vectors and $\alpha = 1$ in (3.106). The number of hidden neurons was chosen equal to 3 and 1, for the state and output equation of the state space model respectively. Using a quasi-Newton method with BFGS updating of the Hessian, 200 iteration steps were done for obtaining a local optimal solution, with non-zero Kalman gain and non-zero bias terms. For this optimal solution, the error on the training set decreased to $9.2902\,e-03$, and to $6.3738\,e-03$ for the test set. The time needed to obtain this local optimal solution is in the order of magnitude of minutes on a DEC 5000/240 workstation. The simulation results are shown in Figure 3.9. Correlation test according to (3.81) were acceptable.

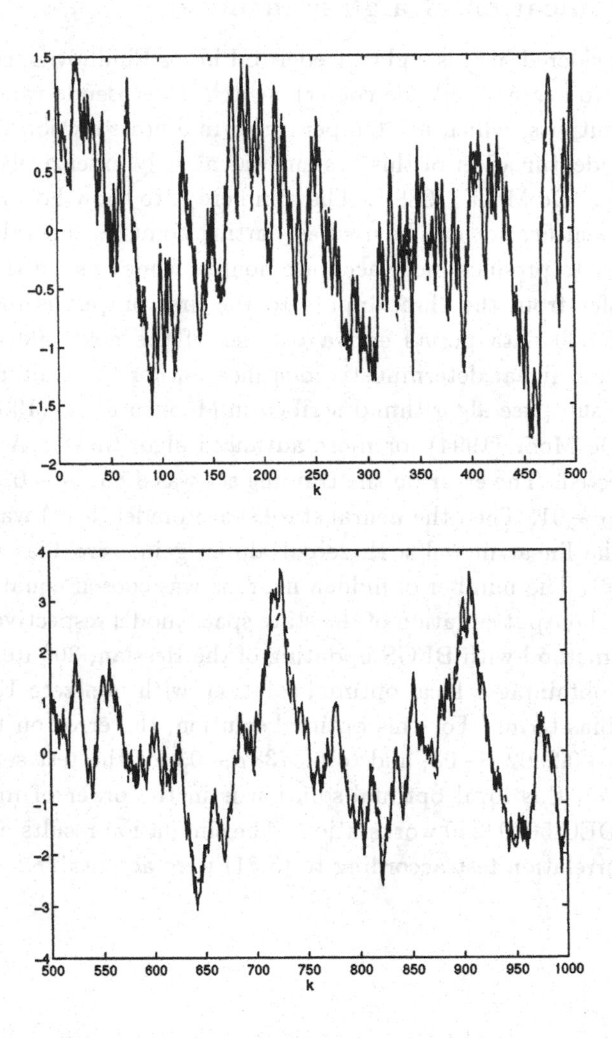


Figure 3.9: *Nonlinear system identification of a glass furnace using neural state space models. Shown on these Figures is one of the temperatures (scaled), measured at a cross section of the oven, which is influenced by two heating and one cooling input. A linear model was used as starting point for the optimization. (Top) result for the training set; (Bottom) result for the test set (fresh data). In both cases: full line = measured output, dashed line = output of neural state space model.*

### 3.7.4 Identifying $n$-double scrolls

In Appendix A, Chua's circuit and its generalization for generating $n$-double scrolls are discussed. We take the same parameter values here. Given the trajectory of the double scroll attractor, obtained by means of a trapezoidal integration rule with constant step length 0.05, we will show that it is possible to emulate these dynamics by means of a simple recurrent neural network emulator.

The recurrent neural network model that we identify here is the neural state space model

$$\hat{x}_{k+1} = W_A \tanh(V_A \hat{x}_k), \qquad \hat{x}_0 = x_0 \tag{3.116}$$

where $\hat{x}_k \in \mathbb{R}^3$. This vector has to track the double scroll trajectory $(x(t), y(t), z(t))$ of (A.1) at the sampling points. We take a small network with 3 hidden neurons. Hence according to the prediction error method one has to minimize

$$\min_{\theta} J = \frac{1}{2N} \sum_{k=1}^{N} (x_k - \hat{x}_k(\theta))^T \, (x_k - \hat{x}_k(\theta)). \tag{3.117}$$

with $\theta = [W_A(:); V_A(:)]$. The gradient of the cost function follows from dynamic backpropagation. Starting from random initial parameter vectors it turned out that this batch-learning of the trajectory by means of a gradient based multi-start local optimization method was impossible (or at least extremely hard). In fact this doesn't come as a surprise, because it is well-known that learning long-term dependencies with gradient descent is difficult (see e.g. Bengio *et al.* (1994)). The following approach on the other hand was successful: the data set was split into 'packets' of increasing dimension. The procedure is:

- Generate random $\theta_0$ and put $a := 1$.
- Do while $a < a_{final}$
  - $\theta^* = \arg\min_{\theta} J(\theta) = \frac{1}{2a\Delta} \sum_{k=1}^{a\Delta} l(\epsilon_k)$
  - $\theta_0 := \theta^*$ and $a := a + 1$

  End

Here $\theta_0$ is the starting point for the local optimization scheme with local optimal solution $\theta^*$. Hence the time horizon $a\Delta$ is increased until all the $N$ data points are consumed. The idea of this packets strategy is that one doesn't start learning new things before one has memorized the previous parts. The danger for overfitting is not high because of the small scale network.

In order to identify the double scroll we took $\Delta = 50$, $a_{final} = 20$. The following neural state space model behaves like the double scroll:

$$W_A = \begin{bmatrix} 3.191701795026490e+00 & -3.961031505875602e+00 & -2.544300729387972e+00 \\ 6.302937463967251e-01 & 2.746315947131907e+00 & 8.024248038305574e-01 \\ -1.411085817901605e+00 & 8.436161546347900e+00 & 3.174868294146957e+00 \end{bmatrix}$$

$$V_A = \begin{bmatrix} -2.446514424620466e-01 & 1.557093499163188e+00 & -6.192223155626144e-01 \\ 4.935534636051888e-01 & -9.111907179654775e-01 & 6.785493890595847e-01 \\ -1.711046461663226e+00 & 3.794433297552783e+00 & -2.105183558255941e+00 \end{bmatrix} \quad (3.118)$$

for initial state $x_0 = [0.9365; -0.0610; -0.1889]$. The simulation results are shown in Figure 3.10 and 3.11. The first 1000 data points were used for the training set. Due to the chaotic nature of the system the error becomes larger behind the vertical line for the test data.

Identifying networks in the series-parallel mode is much easier. We illustrate this by means of the 2-double scroll attractor (see also Appendix A). Let us define the first state variable as the output of the system and take the predictor

$$\hat{y}_k = w^T \tanh(V[y_{k-1}; y_{k-2}; y_{k-3}] + \beta). \quad (3.119)$$

Remark that according to Takens' embedding theorem one should take an embedding dimension equal to $2n+1 = 7$, with $n = 3$ the dimension of the state space (see e.g. Casdagli (1989)). This embedding dimension is larger than the one proposed in (3.119). The training set consist of the first 2000 data points. A neural network with 6 hidden neurons was selected. Twenty runs with random starting points (normal distribution with zero mean and standard deviation 0.1) for the parameter vector were done. A prediction error algorithm according to Section 3.3 was applied. 500 iteration steps were taken for the optimization. The model with a minimal prediction error $4.8119\,e-04$ on the training set is shown in Figure 3.12. In this series-parallel mode the difference between the two signals is almost invisible, which is in sharp contrast to identification with recurrent networks for (3.116). This is due to the fact that the true outputs are used as the input of the neural network model. The computational cost on a DEC 5000/240 workstation is in the order of magnitude of minutes for the feedforward network (3.119) and hours for the recurrent network (3.116), which shows again that identification for recurrent networks is much harder than for feedforward networks.

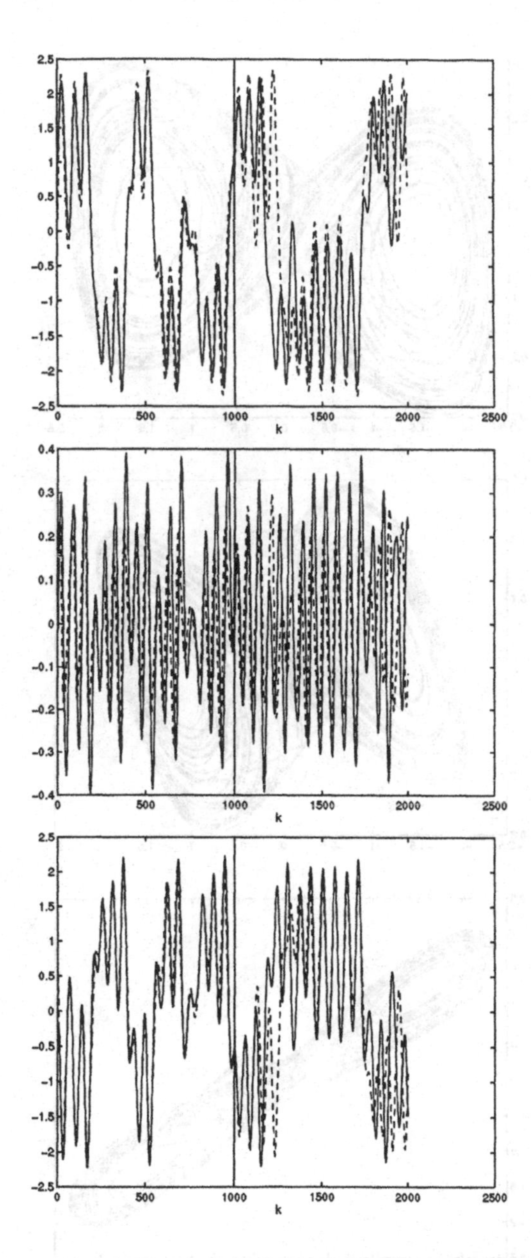


Figure 3.10: *A simple recurrent neural network emulator for Chua's double scroll using neural state space models. First, second and third state variables of Chua's circuit through time k are shown at the top, middle and bottom respectively. The first 1000 data points were used for training.*

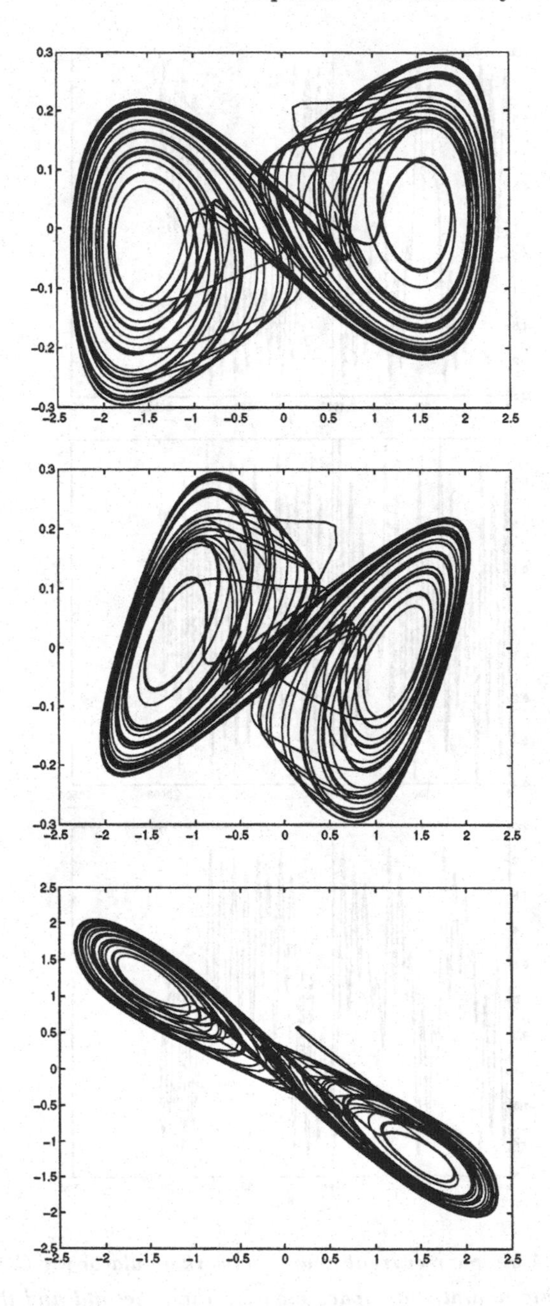


Figure 3.11: *Double scroll attractor, reconstructed by the third order autonomous neural state space model with three hidden neurons.(Top)* $(x_k, y_k)$*; (Middle)* $(z_k, y_k)$*; (Bottom)* $(x_k, z_k)$*.*

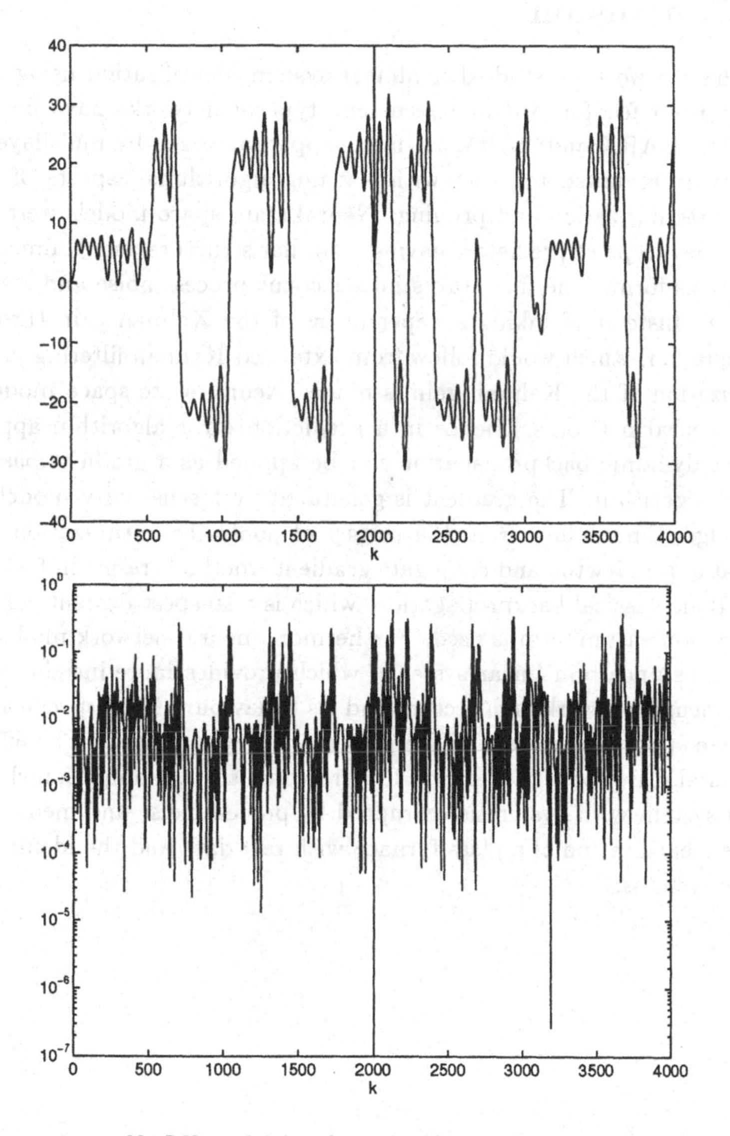


Figure 3.12: *NARX model for the 2-double scroll. An autonomous system of order three was defined. The output is the first state variable of the 2-double scroll circuit. (Top) The first 2000 data points on the plot were used for training. True outputs are in full line, estimated outputs in dashed line. In the series-parallel mode the difference between these signals is almost invisible. Behind the vertical line the next 2000 test data points are shown, simulated in series-parallel mode; (Bottom) absolute value of the prediction error through time.*

## 3.8 Conclusion

In this Chapter we have studied nonlinear system identification using neural networks. Both feedforward and recurrent type of networks have been discussed. First NARX and NARMAX models, parametrized by multilayer perceptrons, were reviewed together with learning algorithms, aspects of model validation, regularization and pruning. Neural state space models were introduced as a new type of predictor, having a similar structure as a Kalman filter in innovations form. The filter takes into account process noise and measurement noise. Instead of taking a dependency of the Kalman gain through a Riccati equation, which would follow from extended Kalman filtering, a direct parametrization of the Kalman gain is made. Neural state space models are recurrent neural networks. Hence in a prediction error algorithm approach, Narendra's dynamic backpropagation can be applied as a gradient based optimization algorithm. The gradient is generated by a sensitivity model. Fast learning algorithms follow from the theory of nonlinear optimization, which shows that quasi-Newton and conjugate gradients methods result in faster convergence than classical backpropagation, which is a steepest descent algorithm in case no momentum term is used. Furthermore, neural network models were interpreted as uncertain linear systems, which provides more insight into the black box neural network architecture and its behaviour. More specifically, for neural state space models, an LFT representation was derived. The effectiveness of neural state space models was illustrated on some examples, including a nonlinear system with hysteresis corrupted by process noise and measurement noise, the identification of a glass furnace with real data and the identification of chaotic systems.

# Chapter 4

# Neural networks for control

In this Chapter we provide the reader with some background material on neural control strategies and we discuss neural optimal control in more detail. The Chapter is organized as follows. In Section 4.1 the basic principles of existing methods in neural control are presented, including direct and indirect adaptive control, reinforcement learning, neural optimal control, internal model control and model predictive control. In Section 4.2 neural optimal control is discussed with respect to classical theory of nonlinear optimal control. The emphasis in this Section is on the formulation of control problems as parametric optimization problems, where static and dynamic nonlinear controllers are parametrized by multilayer perceptrons. Furthermore an efficient way for including a priori results from linear control theory into the neural controller is highlighted. The latter is illustrated on the examples of swinging up an inverted and double inverted pendulum system. New contributions are stated in Section 4.2.6.

## 4.1 Neural control strategies

### 4.1.1 Direct versus indirect adaptive methods

The difference between direct and indirect methods is usually made in the context of adaptive control. In indirect methods, first system identification from I/O measurements on the plant is done and then a controller is learned based upon the identified model. In direct methods the controller is learned directly without having a model from the plant. In neural adaptive control using the backpropagation algorithm the difference becomes clear as follows. Given the control scheme of Figure 4.1, consider a neural controller in order to let the output $y_k$ of the plant track the reference input $d_k$ with cost function

$$\min_{w^l_{ij}} J = \sum_k J_k, \qquad J_k = \frac{1}{2}(d_k - y_k)^2 \tag{4.1}$$

where $w^l_{ij}$ is the $ij$-th interconnection weight at layer $l$ of the neural controller. The neural controller is described by (2.5) with boundary conditions

$$\begin{aligned} x^L_k &= u_k \\ x^0_k &= [d_k; d_{k-1}; ...; d_{k-n_d}; y_k; y_{k-1}; ...; y_{k-n_y}; u_{k-1}; ...; u_{k-n_u}]. \end{aligned}$$

Application of the generalized delta rule gives

$$\left\{ \begin{aligned} \Delta w^l_{ij} &= \eta\, \delta^l_{i,k}\, x^{l-1}_{j,k} \\ \delta^L_{i,k} &= \frac{\partial J_k}{\partial y_k} \frac{\partial y_k}{\partial x^L_{i,k}}\, \sigma'(\xi^L_{i,k}) = (d_k - y_k) \frac{\partial y_k}{\partial x^L_{i,k}}\, \sigma'(\xi^L_{i,k}) \\ \delta^l_{i,k} &= (\textstyle\sum_{r=1}^{N_{l+1}} \delta^{l+1}_{r,k}\, w^{l+1}_{ri})\, \sigma'(\xi^l_{i,k}), \qquad l = 1, ..., L-1. \end{aligned} \right. \tag{4.2}$$

As we have seen in Chapter 2 this algorithm can either be applied in batch mode or on-line. From (4.2) it becomes clear that in order to make backpropagation applicable one needs an expression for

$$\frac{\partial y_k}{\partial x^L_{i,k}}. \tag{4.3}$$

This results in two possible approaches

1. *Indirect neural adaptive control:*
   First an emulator (neural network model) is derived from I/O measurements on the plant. The derivative (4.3) is then computed based upon the emulator. This approach is taken e.g. in Nguyen & Widrow (1990) and has been illustrated on the well known example of backing up a truck.

2. *Direct neural adaptive control:*
   In this case one doesn't use a model for the plant. Instead of (4.3) one takes then

$$\text{sign}(\frac{\partial y_k}{\partial x^L_{i,k}}) \tag{4.4}$$

   such that the error can be backpropagated through the plant. In many cases this is clear from physical insight or can be estimated through some experiments. This procedure was proposed in Psaltis *et al.* (1988), Saerens & Soquet (1991) and Schiffmann & Geffers (1993).

In this book we will mainly focus on indirect model-based neural control schemes.

Finally we have to mention that direct adaptive control methods that make use of radial basis function networks are developed in Sanner & Slotine (1992) and Slotine & Sanner (1993). Principles of feedback linearization were applied to neural adaptive control problems e.g. in Liu & Chen (1993).

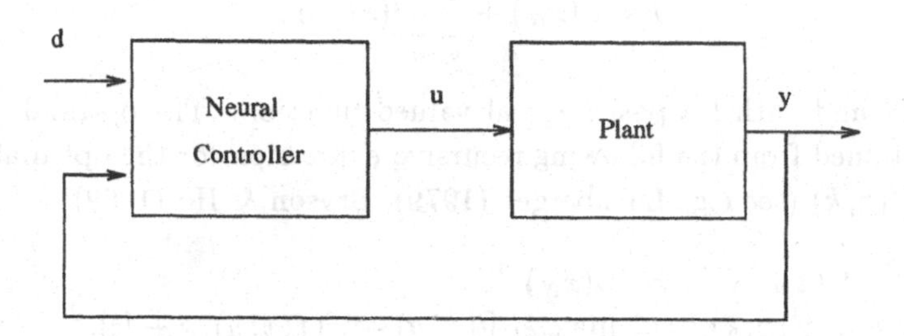


Figure 4.1: *Neural control scheme with input $u$ and output $y$ of a plant and reference input $d$ to be tracked. In an indirect adaptive control scheme, an emulator is first derived from input/output measurements on the plant and the neural controller is designed based upon this model. In direct adaptive control schemes the neural controller is trained without having a model for the plant.*

### 4.1.2 Reinforcement learning

The reinforcement learning method in control is a direct adaptive control strategy. Without having a model for the plant, a so-called critic element evaluates the performance of the control system. The reinforcement learning controller is rewarded or punished depending on the outcome of a number of trials with the system (see Figure 4.2). The reinforcement learning control strategy was first illustrated on the example of broomstick balancing (Barto *et al.* (1983)). In this case the state space was divided into boxes (decoded state space) and through a number of trials a control action is related to each box.

The $Q$-learning method in reinforcement learning is closely related to dynamic programming. Whereas dynamic programming can be applied when a model for the plant is given, $Q$-learning applies to the case where a model is not available and is in fact a direct adaptive optimal control strategy (Sutton *et al.* (1992)). It is well known that Bellman's principle of optimality in dynamic programming states that from any point on an optimal trajectory, the remaining trajectory is optimal for the corresponding problem initiated at that point. Dynamic programming can be formulated as follows. Given a system

$$x_{k+1} = f(x_k, u_k), \qquad x_0 \text{ given} \tag{4.5}$$

with control constraints $u_k \in U$, suppose the objective is to maximize the cost function

$$J = \psi(x_N) + \sum_{k=0}^{N-1} l(x_k, u_k) \tag{4.6}$$

for fixed $N$ and with $l$ a positive real valued function. The optimal solution is then obtained from the following recursive expression for the optimal return function $V(x, k)$ (see e.g. Luenberger (1979), Bryson & Ho (1969))

$$\begin{aligned} V(x_N, N) &= \psi(x_N) \\ V(x, k) &= \max_{u \in U}[l(x, u) + V(f(x, u), k + 1)]. \end{aligned} \tag{4.7}$$

Now in Sutton *et al.* (1992) it is precisely stated that $Q$-learning is an on-line incremental approximation to dynamic programming. Considering a finite state finite action Markov decision problem, the controller observes at each time $k$ the state $x_k$, selects an action $a_k$, receives a reward $r_k$ and observes the next state $x_{k+1}$. The objective is to find a control rule that maximizes at each step the expected discounted sum of future reward

$$E\{\sum_{j=1}^{\infty} \gamma^j r_{k+j}\} \tag{4.8}$$

with discount factor $\gamma$ $(0 < \gamma < 1)$. The basic idea in $Q$-learning is to estimate a real valued function $Q(x, a)$ of state $x$ and action $a$, which is the expected discounted sum of future reward for performing action $a$ in state $x$ and performing optimality thereafter. This function satisfies the recursive relationship

$$Q(x, a) = E\{r_k + \gamma \max_b Q(x_{k+1}, b) | x_k = x, a_k = a\}. \tag{4.9}$$

Given $x_k$, $a_k$, $r_k$, $x_{k+1}$, the $Q$-learning scheme then works with an estimate $\hat{Q}$ that is updated as

$$\hat{Q}(x_k, a_k) := \hat{Q}(x_k, a_k) + \beta_k[r_k + \gamma \max_b \hat{Q}(x_k, b) - \hat{Q}(x_k, a_k)] \tag{4.10}$$

where $\beta_k$ is a gain sequence $(0 < \beta_k < 1)$ and $\hat{Q}(x, a)$ remains unchanged for all pairs $(x, a) \neq (x_k, a_k)$. One disadvantage of this method is that the required memory is proportional to the number of $(x, a)$ pairs which leads to a curse of dimensionality.

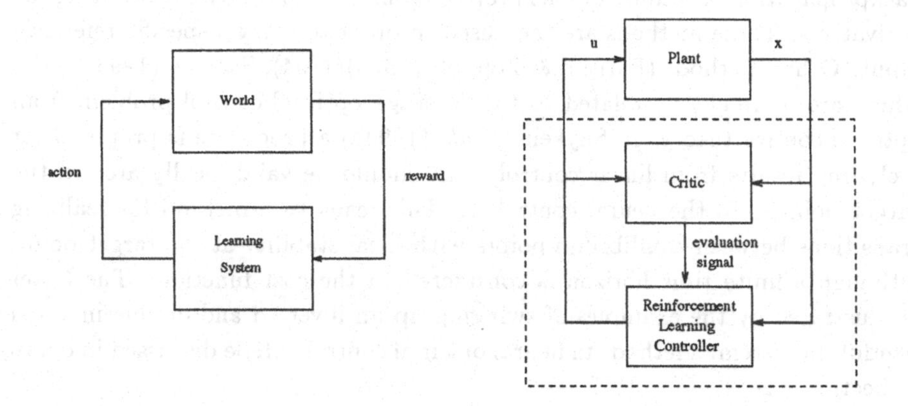


Figure 4.2: *Reinforcement learning control scheme: situations are mapped to actions so as to maximize a scalar reward (reinforcement signal). In the adaptive critic methods the reinforcement learning controller is rewarded or punished by a critic element through a number of trials of the systems. The plant is directly controlled without having a model for the plant.*

### 4.1.3 Neural optimal control

By neural optimal control we mean a model-based neural control strategy, where the controller is parametrized by some neural network architecture and an objective function of interest is minimized (including stabilization and tracking problems). The controller is designed based upon the model (emulator) only. It is usually assumed that the certainty equivalence principle holds which means that the neural controller can be applied to the real plant afterwards instead of to the emulator for the plant. A first example of such a method was presented by Nguyen & Widrow (1990) for backing up a trailer truck. After building an emulator for the trailer truck dynamics (nonlinear system identification), a neural controller was considered and a tracking problem was formulated. The backpropagation method was used in order to minimize the cost function. Backpropagation type algorithms are gradient based methods for optimizing the cost function of interest. However one has to be careful in the computation of the derivatives of the cost function when dynamical systems are involved. As we have discussed in Chapter 2 one has to apply Narendra's backpropagation procedure or backpropagation through time for computing the derivatives. These methods are then used in order to track a specific reference input. Other methods (Parisini & Zoppoli (1991)(1994), Saerens (1993)) exist which are more closely related to the $N$ stage optimal control problem from optimal control theory. In Suykens *et al.* (1994a) a procedure is proposed for including results from linear control, that should be valid locally around the target point, into the neural controller. This leads to a method for realizing transitions between equilibrium points with local stability at the target point, although a finite time horizon is considered in the cost function. The latter is illustrated by the examples of swinging up an inverted and double inverted pendulum system. Methods in neural optimal control will be discussed in detail in Section 4.2.

### 4.1.4 Internal model control and model predictive control

Internal model control and model predictive control are model-based control strategies. In internal model control the control scheme of Figure 4.3 is taken (see Hunt & Sbarbaro (1991), Hunt *et al.* (1992)). The difference between the plant output and the model output is fed back into the control scheme. In this method it is assumed that plant and controller are I/O stable, the inverse of the plant exists and the closed loop system is stable. First a neural emulator M for the plant P is sought, then a neural controller C is considered that is trained to behave as the inverse of the plant and finally the general structure is considered with a filter F.

A model predictive control scheme using neural networks is shown in Figure 4.4 (Sbarbaro-Hofer *et al.* (1993)). The neural network model M provides predictions of the future plant response over a specified time horizon. Predictions supplied by the network are passed to a numerical optimization routine in order to minimize the objective function

$$J = \sum_{j=N_1}^{N_2} (r_{k+j} - \hat{y}_{k+j})^2 + \sum_{j=1}^{N_2} \lambda_j (v_{k+j-1} - v_{k+j-2})^2 \qquad (4.11)$$

subject to the dynamical system model. The neural controller C is then trained to produce the same control output $u$ for given plant output as the outcome $v$ of the optimization routine.

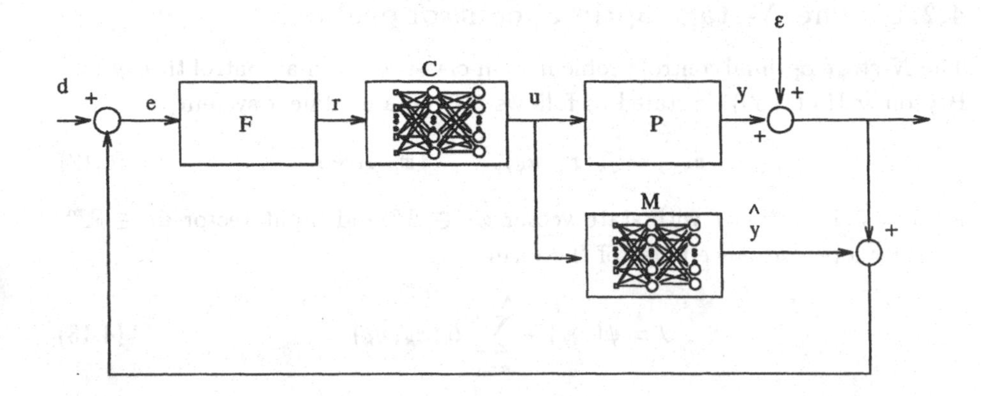


Figure 4.3: *Internal model control scheme. First the plant is identified by means of a neural network emulator M. Second the controller C is trained to represent the inverse of the plant. Furthermore a filter F is considered. P, C and the closed loop system are assumed to be stable and it is supposed that the inverse of P exists.*

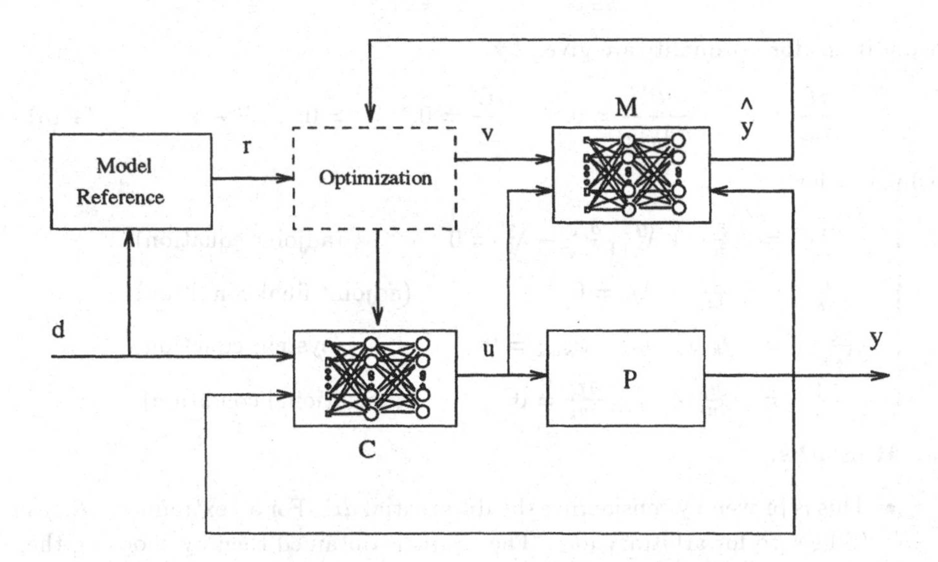


Figure 4.4: *Model predictive control scheme. The neural network model provides predictions for the output of the plant model over a specified time horizon.*

## 4.2 Neural optimal control

### 4.2.1 The $N$-stage optimal control problem

The $N$-stage optimal control problem from classical optimal control theory (see Bryson & Ho (1969)) is stated as follows. Given a nonlinear system

$$x_{k+1} = f_k(x_k, u_k), \qquad x_0 \text{ given} \tag{4.12}$$

for $k = 0, 1, ..., N-1$ with state vector $x_k \in \mathbb{R}^n$ and input vector $u_k \in \mathbb{R}^m$, consider a performance index of the form

$$J = \psi(x_N) + \sum_{k=0}^{N-1} l_k(x_k, u_k) \tag{4.13}$$

which includes the stabilization problem. Let $l_k$ be a positive real valued function. The problem is then to find the sequence $u_k$ that minimizes (or maximizes) $J$. The system equations (4.12) are then adjoined to $J$ with a Lagrange multiplier sequence $\lambda_k$. This yields the Lagrangian

$$\mathcal{L}(x_k, u_k, \lambda_{k+1}) = \psi(x_N) + \sum_{k=0}^{N-1} l_k(x_k, u_k) + \sum_{k=0}^{N-1} \lambda_{k+1}^T (f_k(x_k, u_k) - x_{k+1}). \tag{4.14}$$

Conditions for optimality are given by

$$\frac{\partial \mathcal{L}}{\partial x_k} = 0, \quad \frac{\partial \mathcal{L}}{\partial \lambda_{k+1}} = 0, \quad \frac{\partial \mathcal{L}}{\partial u_k} = 0, \quad k = 0, ..., N-1 \tag{4.15}$$

which yields

$$\left\{ \begin{array}{lcll} \frac{\partial \mathcal{L}}{\partial x_k} & = & \frac{\partial l_k}{\partial x_k} + \lambda_{k+1}^T \frac{\partial f_k}{\partial x_k} - \lambda_k^T = 0 & \text{(adjoint equation)} \\ \frac{\partial \mathcal{L}}{\partial x_N} & = & \frac{\partial \psi}{\partial x_N} - \lambda_N = 0 & \text{(adjoint final condition)} \\ \frac{\partial \mathcal{L}}{\partial \lambda_{k+1}} & = & f_k(x_k, u_k) - x_{k+1} = 0 & \text{(system equation)} \\ \frac{\partial \mathcal{L}}{\partial u_k} & = & \frac{\partial l_k}{\partial u_k} + \lambda_{k+1}^T \frac{\partial f_k}{\partial u_k} = 0 & \text{(variational condition).} \end{array} \right. \tag{4.16}$$

**Remarks.**

- This is proven by considering the differential $d\mathcal{L}$. For an extremum $d\mathcal{L}$ has to be zero for arbitrary $du_k$. The result is obtained then by choosing the multiplier sequence as $\lambda_k^T = \frac{\partial h_k}{\partial x_k}$ with $h_k = l_k(x_k, u_k) + \lambda_{k+1}^T f_k(x_k, u_k)$.
- The resulting problem is a two point boundary value problem. The system equation

$$x_{k+1} = f_k(x_k, u_k)$$

is running forward in time with given $x_0$. The adjoint equation

$$\lambda_k^T = \lambda_{k+1}^T \frac{\partial f_k}{\partial x_k} + \frac{\partial l_k}{\partial x_k}$$

is running backward in time with given $\lambda_N = \frac{\partial \psi}{\partial x_N}$. The continuous time version to (4.16) is known as Pontryagin's maximum principle.

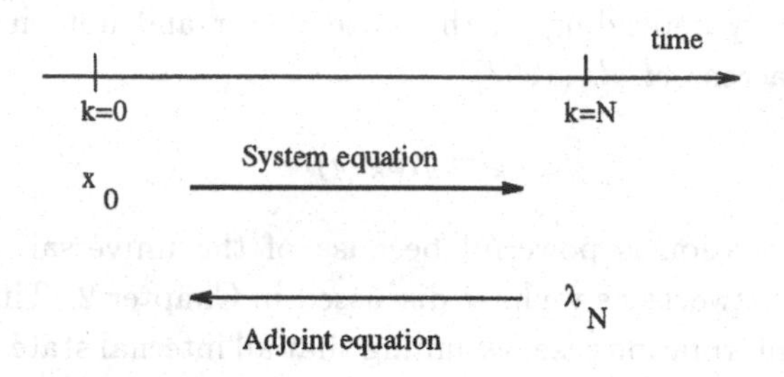


Figure 4.5: *The solution to the N-stage optimal control problem is given by a two-point boundary value problem. The system equation is running forward in time with given initial state $x_0$. The adjoint equation is running backward in time with given final Lagrange multiplier $\lambda_N$.*

- A well-known specific example is the Linear Quadratic problem (LQ). Given the linear system

$$x_{k+1} = A_k x_k + B_k u_k, \qquad x_0 \text{ given} \tag{4.17}$$

the goal is to find the control vector sequence $u_k$, $k = 0, ..., N-1$ such that the following quadratic form is minimized

$$J = \frac{1}{2} x_N^T Q_N x_N + \frac{1}{2} \sum_{k=0}^{N-1} (x_k^T Q_k x_k + u_k^T R_k u_k) \tag{4.18}$$

where $Q_k$, $R_k$ are given positive definite matrices. A solution to the two-point boundary value problem is obtained by imposing a linear dependency between the state variables and the multipliers as

$$\lambda_k = S_k x_k. \tag{4.19}$$

- The conditions (4.15) yield an extremum. To distinguish between a minimum or a maximum one has to consider the second order derivative of (4.13).

### 4.2.2 Neural optimal control: full state information case

### 4.2.3 Stabilization problem: full static state feedback

From (4.16) it is clear that given a nonlinear plant model the optimal control signal will depend both on the state vector and the Lagrange multipliers. Because this is hard to solve in general, it is meaningful to impose a control law which is parametrized by a feedforward neural network with parameter vector $\theta$ and which is only depending on the state vector and not on the Lagrange multiplier (see Saerens *et al.* (1993))

$$u_k = g(x_k; \theta). \tag{4.20}$$

Such a parametrization is powerful because of the universal approximation ability of neural networks as we have discussed in Chapter 2. This corresponds to the full state information case, assuming that all internal states of the system are accessible and measurable. Hence we want to optimize

$$J = \psi(x_N) + \sum_{k=0}^{N-1} l_k(x_k, u_k) \tag{4.21}$$

subject to

1. Dynamical system:
$$x_{k+1} = f_k(x_k, u_k)$$
2. Neural controller:
$$u_k = g(x_k, \theta).$$

The Lagrangian associated to the problem is

$$\begin{aligned} \mathcal{L} = \; & \psi(x_N) + \textstyle\sum_{k=0}^{N-1} l_k(x_k, u_k) + \sum_{k=0}^{N-1} \lambda_{k+1}^T (f_k(x_k, u_k) - x_{k+1}) \\ & + \textstyle\sum_{k=0}^{N-1} \mu_k^T (g(x_k, \theta) - u_k) \end{aligned} \tag{4.22}$$

with multiplier sequences $\{\lambda_k\}$ and $\{\mu_k\}$. The conditions for optimality are

$$\left\{ \begin{aligned} \frac{\partial \mathcal{L}}{\partial x_k} &= \frac{\partial l_k}{\partial x_k} + \lambda_{k+1}^T \frac{\partial f_k}{\partial x_k} - \lambda_k^T + \mu_k^T \frac{\partial g}{\partial x_k} = 0 \\ \frac{\partial \mathcal{L}}{\partial x_N} &= \frac{\partial \psi}{\partial x_N} - \lambda_N = 0 \\ \frac{\partial \mathcal{L}}{\partial \lambda_{k+1}} &= f_k(x_k, u_k) - x_{k+1} = 0 \\ \frac{\partial \mathcal{L}}{\partial u_k} &= \frac{\partial l_k}{\partial u_k} + \lambda_{k+1}^T \frac{\partial f_k}{\partial u_k} - \mu_k^T = 0 \\ \frac{\partial \mathcal{L}}{\partial \mu_k} &= g(x_k, \theta) - u_k = 0. \end{aligned} \right. \tag{4.23}$$

On the other hand let us consider the backpropagation algorithm in order to minimize the performance index (4.21), like for (4.1). Consider the Lagrangian

$$\min_{\theta} \mathcal{L} = \psi(x_N) + \sum_k \mathcal{L}_k \tag{4.24}$$

with

$$\mathcal{L}_k = l_k(x_k, u_k) + \lambda_{k+1}^T (f_k(x_k, u_k) - x_{k+1}) + \mu_k^T (g(x_k, \theta) - u_k) \tag{4.25}$$

where $\theta$ contains the interconnection weights $w_{ij}^l$ of the neural controller. Then the generalized delta rule applied to (4.24) gives

$$\left\{ \begin{array}{rcl} \Delta w_{ij}^l & = & \eta\, \delta_{i,k}^l\, x_{j,k}^{l-1} \\ \delta_{i,k}^L & = & \frac{\partial \mathcal{L}_k}{\partial x_{i,k}^L}\, \sigma'(\xi_{i,k}^L) \\ & = & \mu_{i,k}\, \sigma'(\xi_{i,k}^L) \\ \delta_{i,k}^l & = & (\sum_{r=1}^{N_{l+1}} \delta_{r,k}^{l+1} w_{ri}^{l+1})\, \sigma'(\xi_{i,k}^l), \qquad l = 1, ..., L-1. \end{array} \right. \tag{4.26}$$

Hence the following neural control algorithm is obtained:

- Generate random interconnection weights $w_{ij}^l$ for the neural controller.
- Do until convergence:
  1. *Forward pass:*
     Compute the sequences $\{u_k\}_{k=0}^N$ and $\{x_k\}_{k=0}^N$ from $x_{k+1} = f_k(x_k, u_k)$ and $u_k = g(x_k; \theta)$.
  2. *Backward pass:*
     (a) Compute backward in time ($k = N-1, ..., 0$)

$$\left\{ \begin{array}{rcl} \lambda_N & = & \frac{\partial \psi}{\partial x_N} \\ \mu_k^T & = & \frac{\partial l_k}{\partial u_k} + \lambda_{k+1}^T \frac{\partial f_k}{\partial u_k} \\ \lambda_k^T & = & \frac{\partial l_k}{\partial x_k} + \lambda_{k+1}^T \frac{\partial f_k}{\partial x_k} + \mu_k^T \frac{\partial g}{\partial x_k} \end{array} \right.$$

     in order to obtain the sequence $\{\mu_k\}_{k=0}^{N-1}$.

     (b) Apply the generalized delta rule (2.11) for adapting the weights of the neural controller.

End

### 4.2.4 Tracking problem: the LISP principle

The problem of optimal tracking is considered in Parisini & Zoppoli (1991)(1994). Given the nonlinear system (4.12)

$$x_{k+1} = f_k(x_k, u_k), \qquad x_0 \text{ given}$$

the cost function that is considered is

$$J = \sum_{k=0}^{N-1} [h_k(x_k, u_k) + \rho_{k+1}(\|r_{k+1} - x_{k+1}\|_2)] \tag{4.27}$$

which is in general nonquadratic with $h_k$ and $\rho_{k+1}$ positive nonlinear functions and $r_k \in \mathbb{R}^n$ is a reference state vector. For the Linear Quadratic (LQ) case with system

$$x_{k+1} = Ax_k + Bu_k \tag{4.28}$$

and cost function

$$J = \sum_{k=0}^{N-1} (\|u_k\|_{R_k}^2 + \|r_{k+1} - x_{k+1}\|_{Q_{k+1}}^2) \tag{4.29}$$

the optimal control strategy is basically of the form

$$\left\{ \begin{array}{rcll} u_k & = & -L_k x_k + F_k v_k & k = 0, 1, ..., N-1 \\ v_k & = & G_k v_{k+1} + Q_{k+1} r_{k+1} & k = 0, 1, ..., N-2 \\ v_{N-1} & = & Q_N r_N & \end{array} \right. \tag{4.30}$$

where the matrices $L_k$, $F_k$, $G_k$ can be determined after solving a Riccati equation. The idea applied by Parisini & Zoppoli is the so-called 'LInear Structure Preserving principle' (LISP principle) , which imposes that the optimal control signal to the non-LQ problem (4.27) has to take the form

$$\left\{ \begin{array}{rcll} u_k & = & \gamma(x_k, v_k) & k = 0, 1, ..., N-1 \\ v_k & = & \phi(v_{k+1}, r_{k+1}) & k = 0, 1, ..., N-2 \\ v_{N-1} & = & \varphi(r_N) & \end{array} \right. \tag{4.31}$$

which has the same structure as for the LQ problem, but with nonlinear instead of linear mappings. Multilayer feedforward neural nets are then used in order to parametrize the nonlinear mappings $\gamma$, $\phi$ and $\varphi$. Parisini & Zoppoli applied the generalized delta rule to each of these multilayer feedforward neural networks. The overall algorithm consists of a forward pass for computing the sequences

$x_k$, $u_k$ and a backward pass for computing a multiplier sequence and the delta variables in the backpropagation algorithm.

Due to the generality of the non-LQ problem, it is difficult to apply conventional methods such as the maximum principle or dynamic programming. The method outlined here serves then as an approximate solution by constraining the control strategy to the format (4.31). The method has been applied e.g. in Casalino *et al.* (1993) to a space robot, where the goal is to reach a fixed point in state space while minimizing fuel consumption.

### 4.2.5 Dynamic backpropagation

Assuming the nonlinear plant model is described by a nonlinear state space model

$$\left\{ \begin{array}{lcl} x_{k+1} & = & f(x_k, u_k) \\ \hat{y}_k & = & g(x_k) \end{array} \right. \tag{4.32}$$

consider the nonlinear dynamic output feedback law

$$\left\{ \begin{array}{lcl} z_{k+1} & = & h(z_k, y_k, d_k) \\ u_k & = & s(z_k, y_k, d_k) \end{array} \right. \tag{4.33}$$

with $x_k \in \mathbb{R}^n$, $z_k \in \mathbb{R}^{n_z}$ the state of the plant model and the state of the controller respectively and $u_k \in \mathbb{R}^m$ the input of the plant, $\hat{y}_k \in \mathbb{R}^l$ the output of the plant model and $d_k \in \mathbb{R}^l$ the reference input. The mappings $f(.)$, $g(.)$, $h(.)$ and $s(.)$ are supposed to be continuous and differentiable. In the sequel, parametrizations of these mappings by multilayer feedforward neural networks will be considered. At this point there is however no need to further specify the mappings. The closed loop system for (4.32) and (4.33) is of the form

$$\left\{ \begin{array}{lcl} p_{k+1} & = & \zeta(p_k, d_k; \alpha) \\ \hat{y}_k & = & \chi(p_k, d_k; \beta) \\ u_k & = & \vartheta(p_k, d_k; \gamma). \end{array} \right. \tag{4.34}$$

Suppose the nonlinear controller is parametrized by the parameter vector $\theta_c \in \mathbb{R}^{p_c}$, which contains the elements $\alpha$, $\beta$ and $\gamma$ that correspond to parametrizations for $\zeta$, $\chi$ and $\vartheta$ respectively.

Let us consider now the tracking problem for a specific reference input $d_k$

$$\min_{\theta_c} J(\theta_c) = \frac{1}{2N} \sum_{k=1}^{N} \{[d_k - \hat{y}_k(\theta_c)]^T [d_k - \hat{y}_k(\theta_c)] + \lambda . u_k(\theta_c)^T u_k(\theta_c)\}. \tag{4.35}$$

After parametrizing the model and the controller by neural network architectures the control problem is formulated as a parametric optimization problem.

A gradient based optimization scheme such as a steepest descent, conjugate gradient or quasi-Newton method makes then use of the gradient

$$\frac{\partial J}{\partial \theta_c} = \frac{1}{N} \sum_{k=1}^{N} \{[d_k - \hat{y}_k(\theta_c)]^T (-\frac{\partial \hat{y}_k}{\partial \theta_c}) + \lambda . u_k(\theta_c)^T \frac{\partial u_k}{\partial \theta_c}\} \tag{4.36}$$

where the gradients $\frac{\partial \hat{y}_k}{\partial \theta_c}$ and $\frac{\partial u_k}{\partial \theta_c}$ are the output of the sensitivity model

$$\begin{cases} \frac{\partial p_{k+1}}{\partial \alpha} & = & \frac{\partial \zeta}{\partial p_k} . \frac{\partial p_k}{\partial \alpha} + \frac{\partial \zeta}{\partial \alpha} \\ \frac{\partial \hat{y}_k}{\partial \alpha} & = & \frac{\partial \chi}{\partial p_k} . \frac{\partial p_k}{\partial \alpha} \\ \frac{\partial \hat{y}_k}{\partial \beta} & = & \frac{\partial \chi}{\partial \beta} \\ \frac{\partial u_k}{\partial \gamma} & = & \frac{\partial \vartheta}{\partial \gamma}, \end{cases} \tag{4.37}$$

which is a dynamical system with state vector $\frac{\partial p_k}{\partial \alpha}$, driven by the vectors $\frac{\partial \zeta}{\partial \alpha}$, $\frac{\partial \chi}{\partial \beta}$ and $\frac{\partial \vartheta}{\partial \gamma}$. The Jacobian matrices are $\frac{\partial \zeta}{\partial p_k}$ and $\frac{\partial \chi}{\partial p_k}$.

Dynamic backpropagation as defined in Narendra & Parthasarathy (1991) is the steepest descent algorithm

$$\hat{\theta}_c := \hat{\theta}_c - \eta \frac{\partial J}{\partial \theta_c}. \tag{4.38}$$

The cost function (4.35) corresponds to the off-line (batch) version of the algorithm. The on-line algorithm works basically the same, but a shorter time horizon is taken for the cost function. The choice of this horizon depends upon the specific application.

### 4.2.6 Imposing constraints from linear control theory

Let us consider again the stabilization problem. A problem that arises with the control strategies outlined in Section 4.2.2 is that although it might be possible to find a trajectory that brings the initial state to the final state in $N$ steps, it is not guaranteed that for time $k > N$ the state will remain there even for large $N$. Indeed if the closed loop system for a given neural controller is not locally stabilizing at the target point, the slightest perturbation or disturbance will destabilize the system.

The latter will lead us to a local stability analysis of the closed loop system dynamics at the target point. Based on that analysis it will become clear that there are degrees of freedom left for the neural controller, which means that the neural controller can do more than just local stabilization. This observation yields a procedure for realizing transitions between equilibrium points of a nonlinear plant with local stability at the endpoint, where the linear control design at the target point is incorporated in the neural controller and was introduced in Suykens *et al.* (1994a). For industrial plants this problem of non-local control occurs in switching from one working point to another, which is nowadays basically done by human operators that have the required *'Fingerspitzengefühl'* to successfully accomplish this difficult task. In order to illustrate that neural controllers are able to perform such tasks, the problems of swinging up an inverted and double inverted pendulum are investigated.

#### 4.2.6.1 Static feedback using feedforward nets

Let us consider the following single input continuous time nonlinear system:

$$\begin{cases} \dot{x} &= f(x) + g(x)\,u \\ y &= h(x) \end{cases} \tag{4.39}$$

with state vector $x \in \mathbb{R}^n$, input $u \in \mathbb{R}$, output vector $y \in \mathbb{R}^m$ and $f$, $g$, $h$ vector fields defined on $\mathbb{R}^n$ ($f : \mathbb{R}^n \mapsto \mathbb{R}^n$, $g : \mathbb{R}^n \mapsto \mathbb{R}^n$, $h : \mathbb{R}^n \mapsto \mathbb{R}^m$). Furthermore we suppose that $f(0) = 0$, $h(0) = 0$. If this is not the case a change of coordinates can be applied.

A *static* output feedback law of the form

$$u = \alpha \tanh(w^T \tanh(Vy)) \tag{4.40}$$

is investigated first. The unknown parameter vector of the controller consists of the interconnection matrix $V \in \mathbb{R}^{n_h \times m}$ and vector $w \in \mathbb{R}^{n_h}$ and $\alpha$ is the maximum amplitude of the control signal. The number of hidden neurons is $n_h$.

The closed loop system is then

$$\dot{x} = f(x) + g(x)\,\alpha \tanh(w^T \tanh(Vh(x))). \tag{4.41}$$

In order to derive conditions for local stability at $x = 0$ the Jacobian matrix $J$ is calculated and evaluated at $x = 0$:

$$J(0) = f_0 + g(0)\,\alpha w^T V h_0 \tag{4.42}$$

where $f_x$, $g_x$ and $h_x$ denote the matrices with partial derivatives $[\frac{\partial f_i}{\partial x_j}]$, $[\frac{\partial g_i}{\partial x_j}]$ $(i,j = 1,...,n)$ and $[\frac{\partial h_i}{\partial x_j}]$ $(i = 1,...,m,\ j = 1,...,n)$. The matrices $f_0$ and $h_0$ represent the matrices $f_x$ and $h_x$ evaluated at 0. Local stability around $x = 0$ is guaranteed if the eigenvalues $\lambda_i$ of $J(0)$ are lying in the open left half complex plane

$$\mathrm{Re}\,\{\lambda_i(f_0 + g(0)\,\alpha w^T V\, h_0)\} < 0 \qquad i = 1,...,n. \tag{4.43}$$

It is explained now how the condition (4.43) can be related to linear controller design results and how such results can be incorporated into the neural controller. Linear static output feedback $u = -k^T y$ applied to the linearized system around $x = 0$

$$\begin{cases} \dot{x} &= f_0 \, x + g(0) \, u \\ y &= h_0 \, x \end{cases} \tag{4.44}$$

leads to the closed loop system

$$\dot{x} = (f_0 - g(0) \, k^T \, h_0) \, x \tag{4.45}$$

which is stable if

$$\text{Re} \{ \lambda_i (f_0 - g(0) \, k^T \, h_0) \} < 0 \qquad i = 1, ..., n. \tag{4.46}$$

Comparing (4.43) with (4.46) gives the following set of constraints on the set of weights related to a certain linear controller design with static feedback gain $k$

$$k^T = -\alpha \, w^T V. \tag{4.47}$$

Observe that the interconnection weights $w$ and $V$ are not completely determined by (4.47). A certain degree of freedom is left, which means that additional requirements could be achieved by the neural controller, such as realizing a transition between two equilibrium points of the nonlinear plant, which will be illustrated later on.

**Remarks.**

- It is straightforward to see that the results of one hidden layer (4.47) extend to feedforward neural networks with more than one hidden layer. We can summarize the conditions for local stability at the target equilibrium point as

  1. *One neuron:*
  $$k^T = -\alpha \, w^T \tag{4.48}$$
  2. *One hidden layer:*
  $$k^T = -\alpha \, w^T V$$
  3. *General case of $q$ hidden layers:*
  $$k^T = -\alpha \, w^T V_1 V_2 ... V_q. \tag{4.49}$$

Remark that in the one neuron case the interconnection weights are completely determined by (4.48) for a fixed $\alpha$. This means that linear static state feedback can be placed at the same level as static state feedback with one neuron. In the other cases (4.47)(4.49) the local stability conditions impose a set of constraints on the choice of the interconnection weights of the neural controller.

- In the case of full state feedback we have $h_0 = I_n$ if $y = x$ and the set of constraints on $w, V$ can then be related e.g. to an LQR design (Franklin *et al.* (1990)). Given a linear time-invariant system, the optimal control signal that minimizes the cost function

$$C_{lqr} = \int_0^\infty (x^T Q x + u^T R u)\, dt \tag{4.50}$$

with $Q$ and $R$ given positive definite symmetric matrices, is given by the full static state feedback law $u = -k_{lqr}^T x$, with

$$k_{lqr}^T = R^{-1} g(0)^T P. \tag{4.51}$$

Here $P$ is the stabilizing solution to the matrix algebraic Riccati equation

$$O = P f_0 + f_0^T P - P g(0) R^{-1} g(0)^T P + Q. \tag{4.52}$$

Other techniques like $H_\infty$ control (state feedback case) can be applied here.

#### 4.2.6.2 Dynamic feedback using recurrent nets

Consider again the nonlinear system (4.39) but under the nonlinear dynamic output feedback law

$$\begin{cases} \dot{z} &= \nu(z, y; \theta_1) \\ u &= \rho(z, y; \theta_2) \end{cases} \tag{4.53}$$

with $z \in \mathbb{R}^{n_z}$ the state of the controller, $\theta_1 \in \mathbb{R}^{p_1}$ and $\theta_2 \in \mathbb{R}^{p_2}$ parameter vectors to be determined and $\nu : \mathbb{R}^{n_z} \times \mathbb{R}^m \mapsto \mathbb{R}^{n_z}$, $\rho : \mathbb{R}^{n_z} \times \mathbb{R}^m \mapsto \mathbb{R}$ continuous nonlinear mappings.

Let us parametrize the mappings $\nu$ and $\rho$ by multilayer feedforward neural networks. This gives the following recurrent neural network for the controller

$$\begin{cases} \dot{z} &= W_1 \tanh([V_{11}\ V_{12}] \cdot \begin{bmatrix} z \\ y \end{bmatrix}) \\ u &= \alpha \tanh(w_2^T \tanh([V_{21}\ V_{22}] \cdot \begin{bmatrix} z \\ y \end{bmatrix})) \end{cases} \tag{4.54}$$

with interconnection matrices $W_1 \in \mathbb{R}^{n_z \times n_{h_1}}$, $V_{11} \in \mathbb{R}^{n_{h_1} \times n_z}$, $V_{12} \in \mathbb{R}^{n_{h_1} \times m}$, $V_{21} \in \mathbb{R}^{n_{h_2} \times n_z}$, $V_{22} \in \mathbb{R}^{n_{h_2} \times m}$ and vector $w_2 \in \mathbb{R}^{n_{h_2}}$. The parameter vectors $\theta_1$ and $\theta_2$ in (4.53) contain respectively the elements of $W_1$, $V_{11}$, $V_{12}$ and $\alpha$, $w_2$, $V_{21}$, $V_{22}$.

The closed loop system of the model (4.39) under the feedback law (4.54) is

$$\begin{cases} \dot{z} &= W_1 \tanh([V_{11}\ V_{12}] \begin{bmatrix} z \\ h(x) \end{bmatrix}) \\ \dot{x} &= f(x) + g(x)\, \alpha \tanh(w_2^T \tanh([V_{21}\ V_{22}] \begin{bmatrix} z \\ h(x) \end{bmatrix})). \end{cases} \tag{4.55}$$

Local stability around the target point $x = 0$, $z = 0$ is guaranteed if the $n + n_z$ eigenvalues $\lambda_i$ of the Jacobian $J(0,0)$ are lying in the open left half complex plane:

$$\mathrm{Re}\,\{\lambda_i(J(0,0))\} < 0 \qquad i = 1, ..., n + n_z \tag{4.56}$$

where

$$J(0,0) = \begin{bmatrix} W_1 V_{11} & W_1 V_{12}\, h_0 \\ g(0)\alpha\, w_2^T V_{21} & f_0 + g(0)\alpha\, w_2^T V_{22}\, h_0 \end{bmatrix}. \tag{4.57}$$

Like in the static output feedback case it will be shown how linear controller design results can be included in the neural controller. Linear dynamic output feedback

$$\begin{cases} \dot{z} &= E\,z + F\,y \\ u &= G\,z + H\,y \end{cases} \tag{4.58}$$

applied on the linearized system around $x = 0$

$$\begin{cases} \dot{x} &= A\,x + B\,u \\ y &= C\,x \end{cases} \tag{4.59}$$

with $A = f_0$, $B = g(0)$, $C = h_0$ leads to the closed loop system

$$\begin{cases} \dot{z} &= E\,z + FC\,x \\ \dot{x} &= BG\,z + (A + BHC)\,x \end{cases} \tag{4.60}$$

which is stable if

$$\mathrm{Re}\,\{\lambda_i(\begin{bmatrix} E & FC \\ BG & A + BHC \end{bmatrix})\} < 0 \qquad i = 1, ..., n + n_z. \tag{4.61}$$

Let us assume such a linear stabilizing controller is available, obtained by means of classical, modern or modern robust control techniques, including e.g. *PID*, *LQG*, $H_2$, $H_\infty$ and $\mu$ controllers (see e.g. Åström, Wittenmark (1984); Franklin *et al.* (1990); Maciejowski (1989)). Then taking the interconnection weights of the neural controller (4.54) such that

$$\begin{bmatrix} E & F \\ G & H \end{bmatrix} = \begin{bmatrix} W_1 V_{11} & W_1 V_{12} \\ \alpha\, w_2^T V_{21} & \alpha\, w_2^T V_{22} \end{bmatrix} \tag{4.62}$$

results in a locally stabilizing neural controller. Again the condition (4.62) imposes a set of constraints on the choice of the interconnection weights and some degree of freedom is left.

#### 4.2.6.3 Transition between equilibrium points

In Section 4.2.6.1 and 4.2.6.2 we have derived conditions for local stability at some target equilibrium point of a nonlinear plant controlled by a feedforward or recurrent neural network. From the expressions (4.47) and (4.62) it became clear that the neural controller can do more than just local stabilization, because the imposed set of constraints is underdetermined. It will be shown how these additional degrees of freedom can be exploited in order to realize transitions between equilibrium points with local stability at the target equilibrium point. This is done by formulating a parametric optimization problem in the unknown interconnection weights, constrained by local stability condition at the endpoint.

Let us consider again the static feedback controller (4.40). In order to impose a transition from a given initial state $x(0)$ to a target equilibrium point $x_{eq}$ (we suppose $x_{eq} = 0$ without loss of generality) to the system (4.41) the following optimization problem is formulated

$$\min_{\alpha,w,V} \; C(\alpha, w, V, f, g, h, x(0), \eta, \zeta, T) \tag{4.63}$$

where the cost function $C$ is defined as

$$C = \eta(x(T)) + \int_0^T \zeta(x(t), u(t))\, dt$$

constrained by

1. Local stability at $x = 0$ (4.47).
2. Closed loop system dynamics (4.41).

where e.g. $\eta(x(T)) = \|x(T)\|$ and $\zeta(x(t), u(t)) = x(t)^T Q x(t) + u(t)^T R u(t)$ (quadratic control) or $\zeta = 0$ (terminal control) and $T$ is the time horizon (Bryson, Ho (1969)). Hence (4.63) is a non-local stabilization type of problem.

For the dynamic output feedback case with target point $x = 0, z = 0$ one has the optimization problem

$$\min_{\alpha,W_1,w_2,V_{11},V_{12},V_{21},V_{22}} C(\alpha, W_1, w_2, V_{11}, V_{12}, V_{21}, V_{22}, f, g, h, x(0), z(0), \eta, \zeta, T) \tag{4.64}$$

where the cost function $C$ is defined as

$$C = \eta(x(T), z(T)) + \int_0^T \zeta(x(t), z(t), u(t))\, dt$$

with constraints

1. Local stability at $x = 0,\ z = 0$ (4.62).
2. Closed loop system dynamics (4.55).

The first constraint on (4.63) and (4.64) can be eliminated and the number of unknowns can be reduced. This is done as follows. For the static feedback case, let us assume that the matrix $V$ is square and of full rank. Then $w^T = -\frac{k^T}{\alpha} V^{-1}$ and the optimization problem is in the interconnection matrix $V$ only:

$$\min_{\alpha,V} C(\alpha, V, f, g, h, x(0), \eta, \zeta, T) \tag{4.65}$$

under the constraint of the closed loop dynamics:

$$\dot{x} = f(x) + g(x)\, \alpha \tanh(-\frac{k^T}{\alpha} V^{-1} \tanh(Vh(x))). \tag{4.66}$$

Also for nonsquare matrices $V$ this constraint can be eliminated (see Suykens *et al.* (1994a)).

In the dynamic output feedback case, if one assumes that the matrices $[V_{11}\ V_{12}]$ and $[V_{21}\ V_{22}]$ are square and of full rank, the local stability constraint can be eliminated such that the optimization problem becomes

$$\min_{\alpha,V_{11},V_{12},V_{21},V_{22}} C(\alpha, V_{11}, V_{12}, V_{21}, V_{22}, f, g, h, x(0), z(0), \eta, \zeta, T) \tag{4.67}$$

under the constraint of the closed loop dynamics:

$$\left\{ \begin{array}{rcl} \dot{z} & = & [E\ F]\, [V_{11}\ V_{12}]^{-1} \tanh([V_{11}\ V_{12}] \begin{bmatrix} z \\ h(x) \end{bmatrix}) \\ \dot{x} & = & f(x) + g(x)\, \alpha \tanh(\frac{1}{\alpha}[G\ H]\, [V_{21}\ V_{22}]^{-1} \tanh([V_{21}\ V_{22}] \begin{bmatrix} z \\ h(x) \end{bmatrix})). \end{array} \right. \tag{4.68}$$

**Remarks.**

- It is not guaranteed a priori whether the linearized region will be entered or not. This question of controllability may depend e.g. on the choice of the time horizon $T$. Moreover the region around $x = 0$ where the linearization is valid depends on the interconnection matrices. But once the linearized region is entered the corresponding linear controller takes over, thereby ensuring that the state will remain in this region for all time even if a finite time horizon $T$ is considered for the optimal control problem. Anyhow the local stability condition constrains the search space and guides the learning process for the unknown parameter vector in a good search direction.
- For the linearized region it can be expected that the robustness is comparable to the corresponding linear controller on which the neural controller was designed. Omitting the constraint (4.47)(4.62) would certainly result in a controller that is less robust to perturbations or may be even locally unstable at the target equilibrium point.
- In case a gradient based optimization algorithm is used for minimizing the cost function, Narendra's sensitivity model approach can be used to calculate the derivatives of the objective function with respect to the interconnection weights of the neural controller. But once the local stability constraint has been eliminated, analytic calculation of the gradient becomes difficult due to the inverse of the interconnection matrices. The gradients can then be calculated numerically or non-gradient based optimization algorithms can be used, such as e.g. genetic algorithms which look only at cost function values (survival of the fittest).
- The determination of the number of hidden neurons needed to perform a certain transition for a given system is a trial and error procedure, but we will illustrate by the examples of swinging up an inverted and double inverted pendulum that simple neural nets are capable of performing complex control tasks.
- It is also possible to let the controller learn for several initial states $x(0)$ by defining, for example, a cost function $C$ that is the sum of the cost functions each related to a single initial state.

#### 4.2.6.4 Example: swinging up an inverted pendulum

We will illustrate the principle of imposing a linear controller design constraint here for the example of swinging up an inverted pendulum system. The system is shown in Figure 4.6. A state space model is given by (4.39) with

$$f(x) = \begin{bmatrix} x_2 \\ \frac{\frac{4}{3}mlx_4^2 sinx_3 - \frac{mg}{2}sin(2x_3)}{\frac{4}{3}m_t - mcos^2 x_3} \\ x_4 \\ \frac{m_t g sinx_3 - \frac{ml}{2}x_4^2 sin(2x_3)}{l(\frac{4}{3}m_t - mcos^2 x_3)} \end{bmatrix}, \; g(x) = \begin{bmatrix} 0 \\ \frac{4}{3}\frac{1}{\frac{4}{3}m_t - mcos^2 x_3} \\ 0 \\ -\frac{cosx_3}{l(\frac{4}{3}m_t - mcos^2 x_3)} \end{bmatrix}, \; h(x) = \begin{bmatrix} x_1 \\ x_3 \end{bmatrix}. \tag{4.69}$$

The state variables $x_1, x_2, x_3, x_4$ are respectively position and velocity of the cart, angle of the pole with the vertical and rate of change of the angle. The input signal $u$ is the force applied to the cart's center of mass. The symbols $m, m_t, l, g$ mean respectively mass of the pole, total mass of cart and pole, half pole length and the acceleration due to gravity. The input signal $u$ is constrained by $|u| < \alpha$. In the sequel we take $m = 0.1$, $m_t = 1.1$, $l = 0.5$. Remark that in the autonomous case $x = [0; 0; 0; 0]$ (pole up) and $x = [0; 0; \pi; 0]$ (pole down) are equilibrium points, respectively called $eq^+$ and $eq^-$. The linearized system around the target equilibrium point $eq^+$ is

$$f_0 = \begin{bmatrix} 0 & 1 & 0 & 0 \\ 0 & 0 & -\frac{mg}{\frac{4}{3}m_t - m} & 0 \\ 0 & 0 & 0 & 1 \\ 0 & 0 & \frac{m_t g}{l(\frac{4}{3}m_t - m)} & 0 \end{bmatrix}, \; g(0) = \begin{bmatrix} 0 \\ \frac{4}{3}\frac{1}{\frac{4}{3}m_t - m} \\ 0 \\ -\frac{1}{l(\frac{4}{3}m_t - m)} \end{bmatrix}, \; h_0 = \begin{bmatrix} 1 & 0 & 0 & 0 \\ 0 & 0 & 1 & 0 \end{bmatrix}. \tag{4.70}$$

The control task of broomstick balancing has been solved in the past by many control strategies. In Barto *et al.* (1983) reinforcement learning was applied. Without having a model for the system, the neural controller was able to learn the stabilization task through a number of trials by exploring the search space. Based on the outcome of the trials the controller is adapted until the control task is accomplished. Also by means of backpropagation type of algorithms the stabilization task was learned in an indirect as well as in a direct neural adaptive control approach (see Saerens & Soquet (1991), Saerens *et al.* (1993), Widrow (1987)). Also visual feedback has been applied (Tolat & Widrow (1988)). But at this level there is in fact no need for applying neural control strategies, because linear control strategies such as LQR and $H_\infty$ control are successful on the broomstick balancing as well. Indeed we have seen that the one-neuron case (no hidden layer) is more or less equivalent to the linear control case.

More difficult than the local stabilization problem is the swinging up problem which is a non-local control task where the system has to go from one equilibrium point to another and has to remain locally stabilized. In order to accomplish this control task two types of neural controllers were proposed in Suykens *et al.* (1994a) (see Figure 4.6).

1. *Feedforward neural controller:*
In this case we need full state information, which can be understood from the linear controller design at the endpoint. We take $y = x$ and a neural controller with 4 hidden neurons. Hence $V \in \mathbb{R}^{4\times4}$ and $w \in \mathbb{R}^4$ in (4.40). The linear controller design at $eq^+$ is done by means of LQR with $Q = I_4$ and $R = 0.01$ in (4.50), which gives the static feedback gain $k_{lqr}$ in (4.47). The maximal force $\alpha$ was chosen equal to 10. In addition to (4.63), a constraint $\|V\|_2 \leq \beta$ was imposed in order to prevent the region where the linearization (4.42) is valid to become too small. The optimization problem was solved here with $\beta = 2$, $x(0) = [0; 0; \pi; 0]$, $T = 3$. The closed loop system is simulated by means of a trapezoidal integration rule (see Rice (1983)) with constant step length 0.03. The simulation is done in C code. The cost function was calculated from this simulation result with terminal control ($\zeta = 0$). A multistart local optimization algorithm SQP (Sequential Quadratic Programming) was applied (*constr* of Matlab) with numerical calculation of the gradients. On a DEC 5000/240 workstation it typically takes a few minutes to obtain a solution to the problem. Simulation results are shown in Figure (4.7). Perturbations on the initial state are shown in Figure (4.9). For a constrained nonlinear optimization problem of the general form

$$\min_{x \in \mathbb{R}^p} \varphi(x) \qquad c(x) \leq 0$$

where $\varphi \in \mathbb{R}$ and $c \in \mathbb{R}^q$, the associated Lagrangian function is

$$L(x, \lambda) = \varphi(x) + \sum_{i=1}^{q} \lambda_i c_i(x)$$

In the SQP method a Quadratic Programming (QP) subproblem is solved at each iteration $x_{k+1} = x_k + \alpha_k d_k$ by linearizing the nonlinear constraints

$$\min_{x \in \mathbb{R}^p} \frac{1}{2} d^T H_k d + \nabla\varphi(x_k)^T d$$

with Hessian update using the BFGS formula (Broyden, Fletcher, Goldfarb, Shanno)

$$H_{k+1} = H_k + \frac{q_k q_k^T}{q_k^T s_k} - \frac{H_k^T H_k}{s_k^T H_k s_k}$$

where $s_k = x_{k+1} - x_k$ and $q_k = \nabla\varphi(x_{k+1}) + \sum_{i=1}^{q} \lambda_i c_i(x_{k+1}) - (\nabla\varphi(x_k) + \sum_{i=1}^{q} \lambda_i c_i(x_k))$ (see Fletcher (1987), Gill *et al.* (1981).

Besides SQP, also global optimization algorithms were applied to this problem. In Thierens, Suykens *et al.* (1993) a genetic algorithm was successfully applied. Another global optimization method by means of a Fokker-Planck Learning Machine (Suykens & Vandewalle (1994i)) was applied and will be explained in detail in Appendix B. The latter has been shown to be more effective than multistart SQP.

2. *Recurrent neural controller:*
A recurrent neural network with $W_1 \in \mathbb{R}^{4\times 6}$, $w_2 \in \mathbb{R}^6$, $V_{11} \in \mathbb{R}^{6\times 4}$, $V_{12} \in \mathbb{R}^{6\times 2}$, $V_{21} \in \mathbb{R}^{6\times 4}$, $V_{22} \in \mathbb{R}^{6\times 2}$ ($n_{h_1} = n_{h_2} = 6$) and $y = [x_1\ x_3]^T$ was taken in (4.55). At the target point an LQG design was done resulting in the locally stabilizing dynamic output feedback law $\{E, F, G, H\}$ in (4.62). The maximal force was chosen by $\alpha = 10$. Again in order to have a sufficiently large region where the linearization is valid additional constraints $\|[V_{11}\ V_{12}]\|_2 \leq \beta$, $\|[V_{21}\ V_{22}]\|_2 \leq \beta$ are imposed to (4.57). Here $\beta = 2$, $x(0) = [0; 0; \pi; 0]$, $z(0) = 0$, $T = 10$ was chosen. The other parameters were the same as in the static feedback case. A trapezoidal integration rule with constant step length 0.01 was used. The cost function $C$ in (4.64) was calculated from the simulation result with terminal control ($\zeta = 0$). A locally optimal solution by means of SQP in shown in Figure 4.8. The recurrent neural controller generalizes much better than the feedforward controller with respect to new initial states (see Figure 4.9). This is probably due to the internal memory of the recurrent neural controller.

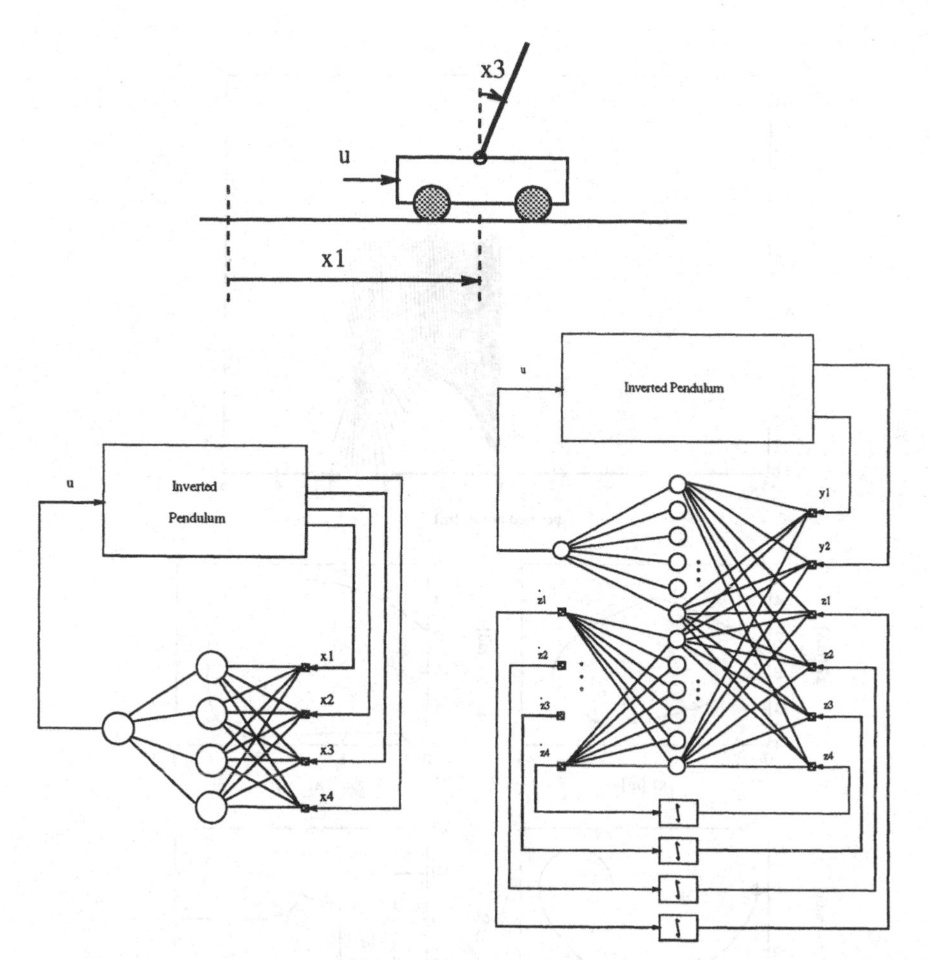


Figure 4.6: *Inverted pendulum system with feedforward and recurrent neural controllers in order to perform the swinging up problem together with local stabilization at the target equilibrium point (upright position). The state variables of the system are:* $x_1$ *the position of the cart,* $x_3$ *the angle of the pole,* $u$ *the force applied to the cart and the velocities* $x_2 = \dot{x}_1$, $x_4 = \dot{x}_3$. *Note that there are two equilibrium positions, the one up which is unstable and the one down, which is stable. The control strategy shown at the left consists of a state feedback controller which manipulates the input (the force applied to the system). The controller at the right is itself a dynamical system (recurrent neural net) using two output measurements: the position of the cart and the angle of the pole. Remark that the neural controllers are relatively simple for accomplishing such a difficult control task.*

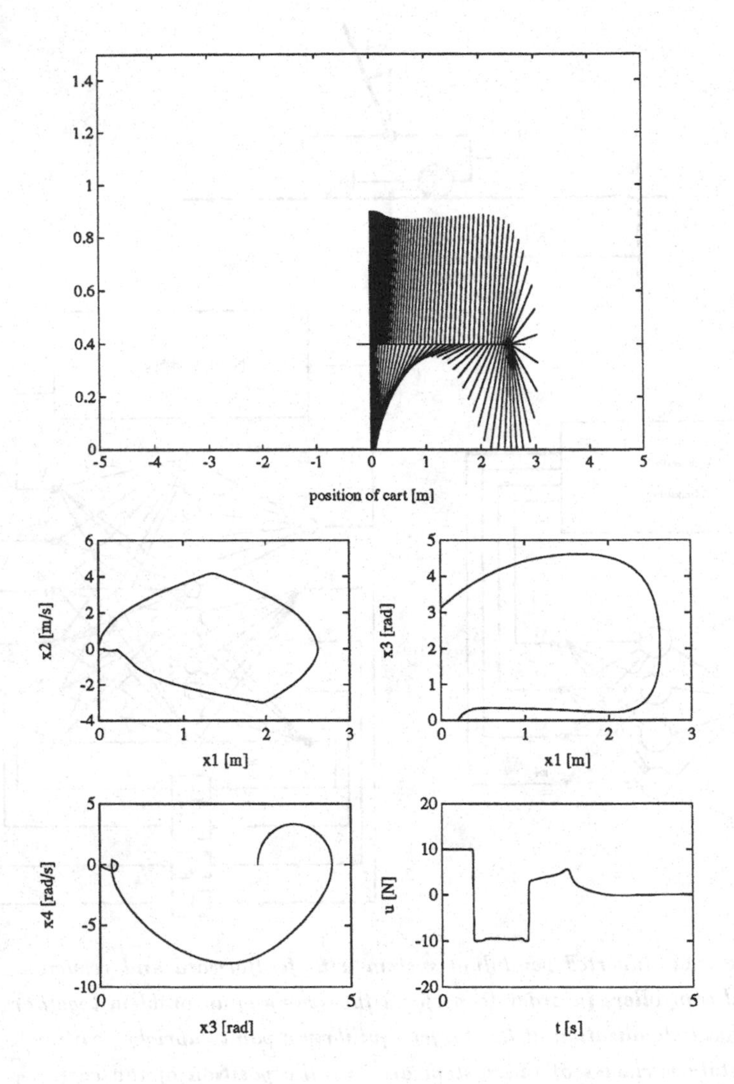


Figure 4.7: *Simulation results of the swinging up problem by means of the feedforward neural controller: The top figure shows the time evolution of the pole. First the cart moves to the right and the pole is swinging behind. Then suddenly the cart moves to the left, an action which swings the pole up. After swinging up, the pole is locally stabilized at its upright position and the neural controller is acting then like an LQR controller. The other figures show the behaviour of the closed loop system in the 4 dimensional state space and the last figure contains the control signal (the applied force) with respect to time.*

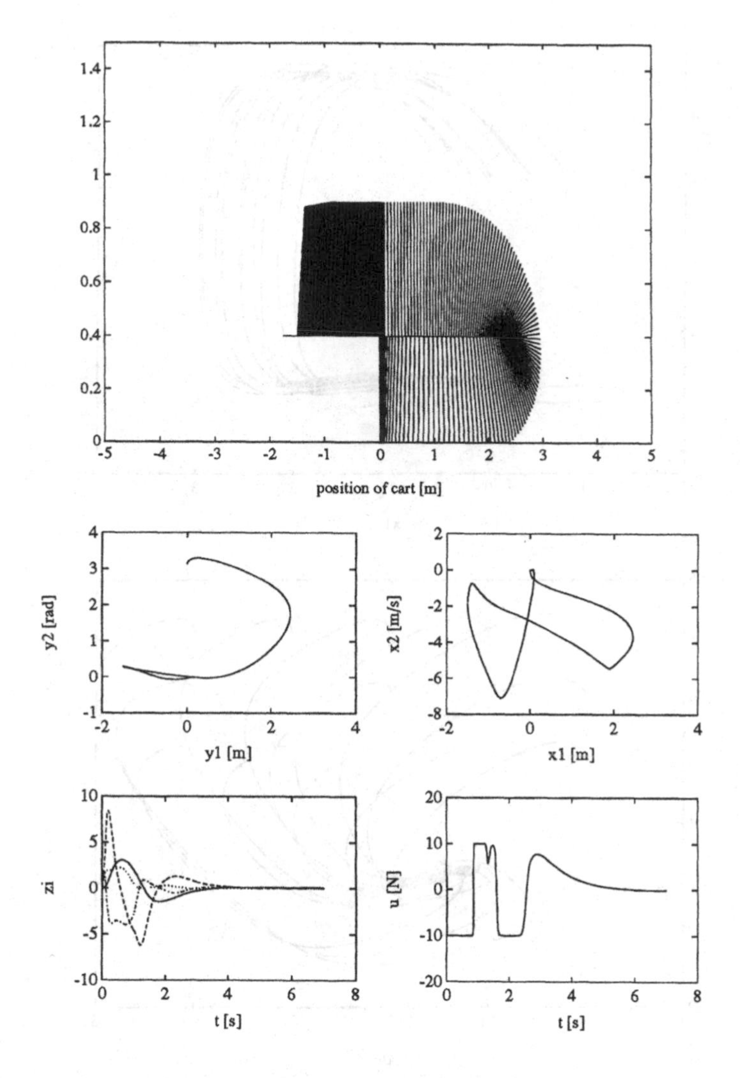


Figure 4.8: *Simulation results of the swinging up problem by means of the recurrent neural controller: The top figure shows the time evolution of the pole. After swinging up the pole is locally stabilized at its upright position. The neural controller is acting then as an LQG controller. The other figures show the behaviour of the closed loop system and the last figure contains the control signal (the applied force) with respect to time. Here we have initial state* $x(0) = [0\ 0\ \pi\ 0]^T$, $z = 0$ *and target equilibrium point* $x = 0$, $z = 0$. *Plot of* $z_i(t)$: $z_1$ *(solid line),* $z_2$ *(dashed line),* $z_3$ *(dotted line),* $z_4$ *(dashdot line).*

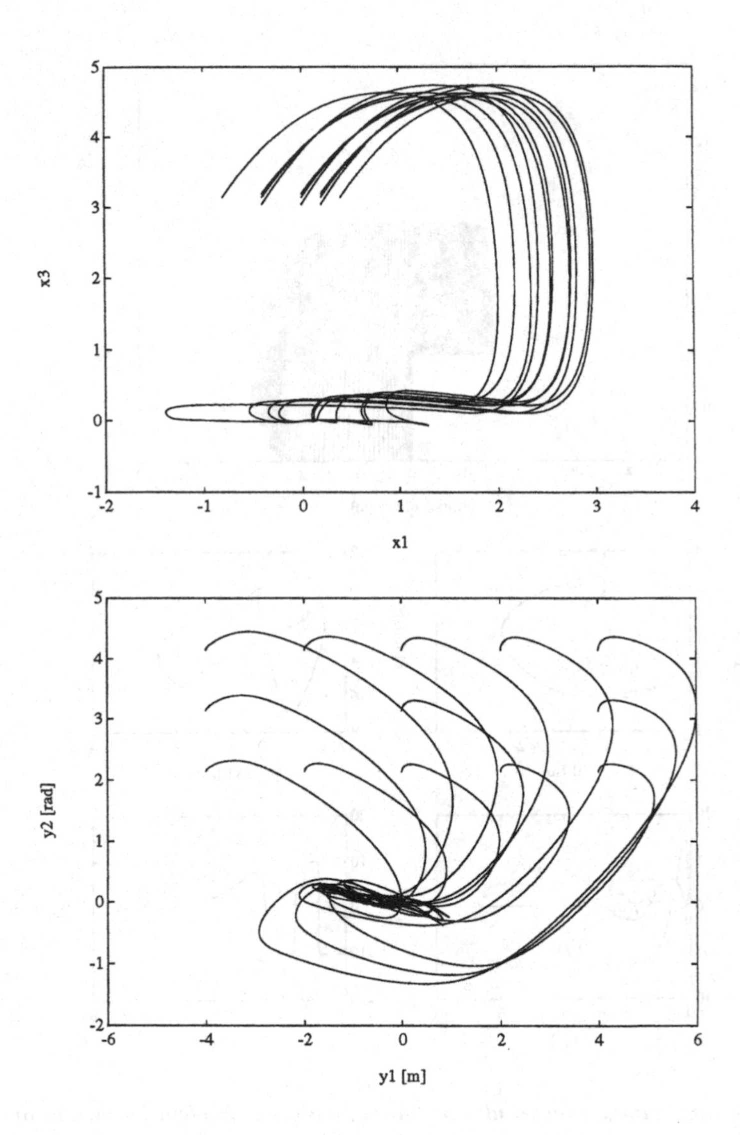


Figure 4.9: *Behaviour of the neural controlled inverted pendulum systems of Figure 4.7 and Figure 4.8 for perturbed initial states* $x(0)$*. Note that* $y_1 = x_1$ *and* $y_2 = x_3$*. A comparison is made between the feedforward neural controller (Top) and the recurrent neural controller (Bottom). All trajectories shown on the Figures are going to the origin. The recurrent neural controller generalizes much better towards initial states which were not taken into account in the learning process for performing the swinging up control task.*

#### 4.2.6.5 Example: swinging up a double inverted pendulum

A second and more difficult example is swinging up a double inverted pendulum with local stabilization at the endpoint. This problem was solved in Suykens *et al.* (1993c). The double inverted pendulum is described by the following standard state space model in robotic terminology (see Larcombe (1991))

$$M(x_p, \theta_1, \theta_2)\ddot{d} + V(x_p, \theta_1, \theta_2, \dot{x}_p, \dot{\theta}_1, \dot{\theta}_2) + G(x_p, \theta_1, \theta_2) = F_e \tag{4.71}$$

with

$$M = \begin{bmatrix} m_c + m_1 + m_2 & (\frac{m_1}{2} + m_2)l_1 cos\theta_1 & \frac{m_2}{2} l_2 cos\theta_2 \\ (\frac{m_1}{2} + m_2)l_1 cos\theta_1 & (\frac{m_1}{3} + m_2)l_1^2 & \frac{m_2}{2} l_1 l_2 cos(\theta_2 - \theta_1) \\ \frac{m_2}{2} l_2 cos\theta_2 & \frac{m_2}{2} l_1 l_2 cos(\theta_2 - \theta_1) & \frac{m_2}{3} l_2^2 \end{bmatrix} \tag{4.72}$$

and

$$V = \begin{bmatrix} (\frac{m_1}{2} + m_2)l_1 sin\theta_1 \dot{\theta_1}^2 \\ \frac{m_2}{2} l_1 l_2 sin(\theta_2 - \theta_1)\theta_2 \\ -\frac{m_2}{2} l_1 l_2 sin(\theta_2 - \theta_1)\dot{\theta_1}^2 \end{bmatrix}, G = \begin{bmatrix} 0 \\ (\frac{m_1}{2} + m_2)l_1 g sin\theta_1 \\ \frac{m_2}{2} l_2 g sin\theta_2 \end{bmatrix}.$$

The system consists of two ideal links with length $l_1, l_2$ and masses $m_1, m_2$, each joint a frictionless revolute hinge. The mass of the cart is $m_c$. The variables $x_p$, $\theta_i (i = 1, 2)$ are respectively equal to the distance of the cart from some fixed reference point and the angle of the $i$-th link against a fixed vertical, measured in clockwise direction. The state vector of the system related to the state space model (4.39) is $x = [x_p; \dot{x}_p; \theta_1; \dot{\theta}_1; \theta_2; \dot{\theta}_2]$ and $d = [x_p; \theta_1; \theta_2]$. The force applied to the cart is equal to $F$. The positive definite matrix $M$ is known as the mass or inertia matrix, the vector $V$ contains centrifugal and Coriolis force terms, $G$ holds gravitational terms and $F_e = [F; 0; 0]$ is an external force vector.

The control task that we consider here is the transition from the initial state $x(0) = [0; 0; \pi; 0; \pi; 0]$ to $x_{eq} = 0$ (swinging up problem) with local stabilization at the endpoint. We take a full static state feedback controller parametrized by a feedforward neural network consisting of 6 hidden neurons. The control signal $u$, which is the force $F$ applied to the cart is then

$$u = \alpha \tanh(w^T \tanh(Vx)) \tag{4.73}$$

with $\alpha$ the maximal force applied to the cart. First a linear controller design is done around the target equilibrium point using LQR with $Q = I_6$ and $R = 1$ with system parameters $m_c = 1$, $m_1 = m_2 = 0.1$, $l_1 = l_2 = 0.5$. For the controller a maximal force $\alpha = 40$ was taken. For the nonlinear optimization problem (4.63) $\eta = \|x(T)\|_2$, $\zeta = 0$ (terminal control), $T = 2$ were chosen and $||V||_2 \leq 5$ was imposed in order to create a sufficiently large region where the linearization is valid. The time horizon $T$ was long enough to let the state entering the linearized region around $x_{eq}$. The closed loop system was simulated in C using a trapezoidal integration rule in the time interval $[0, 2]$ with constant step length equal to 0.002. The nonlinear optimization problem was solved using multistart SQP. The starting points for $V$ were chosen randomly according the a Gaussian distribution with zero mean and variance 0.1. The simulation results are shown in Figure 4.10 and Figure 4.11.

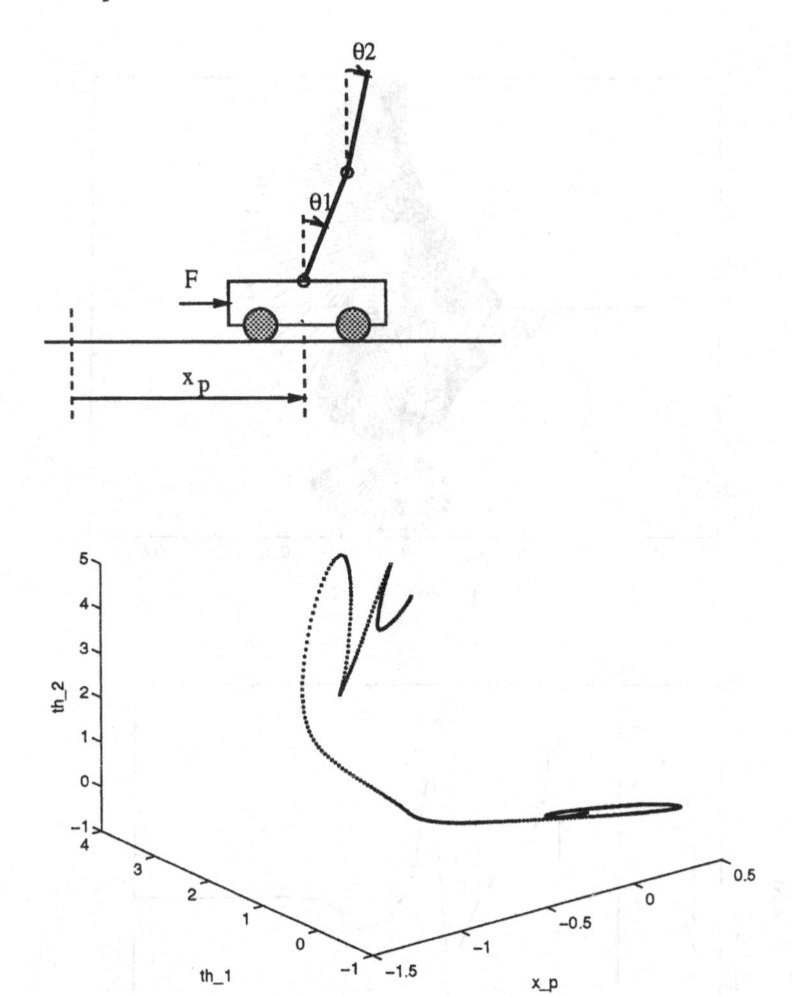


Figure 4.10: *Simulation result to the swinging up problem for a double inverted pendulum using a feedforward neural net controller, consisting of one hidden layer with 6 hidden neurons and applying full state feedback. The state vector is* $x = [x_p; \dot{x}_p; \theta_1; \dot{\theta}_1; \theta_2; \dot{\theta}_2]$. *The control signal is the force* $F$ *applied to the cart. An LQR design was done around the target equilibrium point. The initial state is* $x(0) = [0; 0; \pi; 0; \pi; 0]$ *and the desired final state is* $x_{eq} = 0$. *The Figure at the bottom shows a three-dimensional view of the time trajectory of the vector* $(x_p, \theta_1, \theta_2)$ *of the six dimensional state space.*

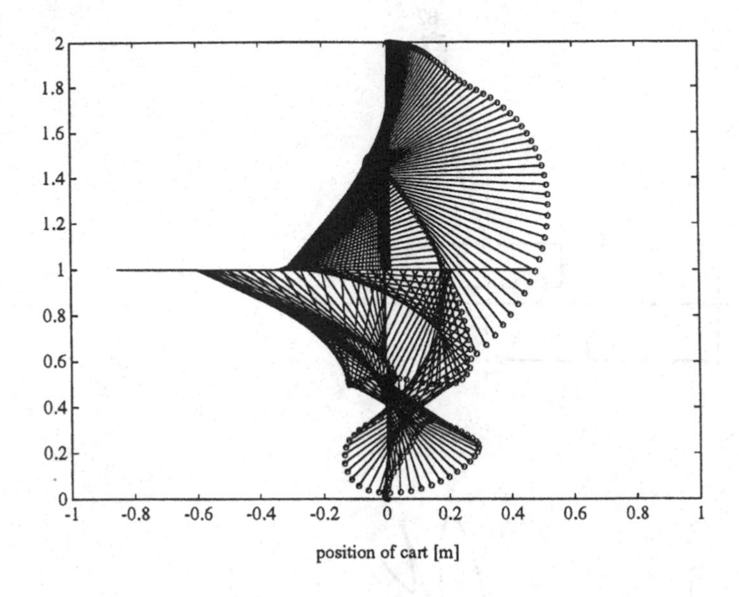


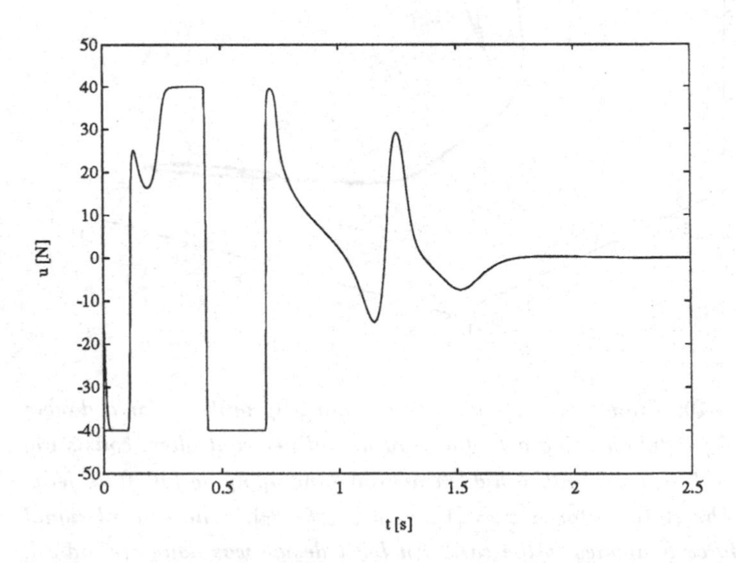


Figure 4.11: *Solution to the swinging up problem of a double inverted pendulum of the previous Figure. (Top) Simulation result, visualized by showing positions of the two poles through time. (Bottom) Force applied to the cart: first a high force is needed in order to swing up the poles and after approximately* 1.5*s the neural controller acts as a classical LQR controller, which locally stabilizes the system at the target point with a smaller force.*

## 4.3 Conclusion

In neural control both model based and non-model based approaches exist. In neural adaptive control this difference is revealed by direct versus indirect methods. In indirect adaptive control an emulator (model) for the plant is needed in order to make backpropagation through the plant applicable. A main difference between neural control and classical linear control is that models and/or controllers are parametrized by multilayer neural networks instead of by linear mappings. Because of their universal approximation ability, this may lead to more accurate model descriptions and to more performant controllers. However this universality is in fact both a strength and a weakness, because their nonlinearity makes them also much harder to analyse. Hence a better understanding of internal and closed loop stability of neural controllers is needed, in order to obtain reliable controller design procedures. This is the topic of the next Chapter.

This Chapter was mainly devoted to neural optimal control, which was developed starting from the basic roots of classical optimal control theory. Suboptimal solutions have been obtained by parametrizing unknown nonlinear functions through multilayer perceptrons, which turns the original problem formulation into parametric optimization problems. The effectiveness of this approach is illustrated on some highly complicated nonlinear control problems of swinging up an inverted and double inverted pendulum. On these examples it is shown that small scale neural networks are able to perform such difficult control tasks. Moreover procedures for including linear controller design results into the neural controller were explained. Linearizing the neural control system at the target equilibrium shows immediately that neural controllers have additional degrees of freedom with respect to linear controllers. This means that neural control can do 'more' than just local stabilization. On the swinging up problems this was illustrated by realizing a transition between two equilibrium points with local stability at the endpoint. Like in nonlinear system identification by means of recurrent neural networks (such as neural state space models), dynamic backpropagation and backpropagation through time play an important role in case the parametric optimization problems are solved by means of gradient based optimization algorithms.

# Chapter 5

# $\mathrm{NL}_q$ Theory

In this Chapter we develop a model based neural control framework which consists of neural state space models and neural state space controllers. Like in modern (robust) control theory standard plant forms are considered. In order to analyse and synthesize neural controllers within this framework, the so-called $\mathrm{NL}_q$ system form is introduced. $\mathrm{NL}_q$s represent a large class of nonlinear dynamical systems in state space form and contain a number of $q$ layers of an alternating sequence of linear and static nonlinear operators that satisfy a sector condition. All system descriptions are transformed into this $\mathrm{NL}_q$ form and sufficient conditions for global asymptotic stability, input/output stability with finite $L_2$-gain and robust performance are derived. It turns out that $\mathrm{NL}_q$s have a unifying nature, in the sense that many problems arising in neural networks, systems and control can be considered as special cases. Moreover, certain results in $H_\infty$ and $\mu$ control theory can be interpreted as special cases of $\mathrm{NL}_q$ theory. Examples show that following principles from $\mathrm{NL}_q$ theory, stabilization and control of several types of nonlinear systems are possible, including mastering chaos.

This Chapter is organized as follows. In Section 5.1 the model based neural control framework with neural state space models and controllers is presented, according to the concepts of modern control theory. In Section 5.2 $\mathrm{NL}_q$ systems are introduced. In Section 5.3 and 5.4 we derive sufficient conditions for global asymptotic stability and I/O stability with finite $L_2$-gain for $\mathrm{NL}_q$s. The connection between internal and I/O stability is revealed through the concept of dissipativity. In Section 5.5 we explain in what sense $\mathrm{NL}_q$ theory is an extension to $\mu$ theory in robust (linear) control. This is done by considering perturbed $\mathrm{NL}_q$s. In Section 5.6 the sufficient stability criteria are formulated as Linear Matrix Inequalities (LMIs). It is well-known that these correspond to convex optimization problems. In Section 5.7 neural control design within $NL_q$ theory is discussed, which finally leads to a modified dynamic backpropagation algorithm in order to assess closed loop stability. Examples on neural control design are given in Section 5.8, including stabilization and tracking problems of nonlinear systems that are globally asymptotically stable, have multiple equilibria, are periodic or even chaotic. Finally in Section 5.9 recurrent neural networks such as generalized cellular neural networks and locally recurrent globally feedforward networks are represented as $NL_q$s.

## 5.1 A neural state space model framework for neural control design

The framework for neural control design that is outlined here is *indirect* in the sense that it is assumed that given measured input/output data on a plant, first nonlinear system identification is done and the control design will be based on a model (emulator) for the plant.

Once a controller is designed it is assumed that the certainty equivalence principle holds which means that the controller can be applied to the plant itself instead of to the model. Consider a nonlinear plant, described in state space form as

$$\begin{cases} x_{k+1} = f_0(x_k, u_k) + \varphi_k \\ y_k = g_0(x_k, u_k) + \psi_k \end{cases} \tag{5.1}$$

with $f_0(.)$ and $g_0(.)$ continuous nonlinear mappings, $u_k \in \mathbb{R}^m$ the input vector, $y_k \in \mathbb{R}^l$ the output vector, $x_k \in \mathbb{R}^n$ the state vector and $\varphi_k \in \mathbb{R}^n$, $\psi_k \in \mathbb{R}^l$ respectively process noise and measurement noise, assumed to be zero mean white Gaussian with covariance matrices

$$E\{\begin{bmatrix} \varphi_k \\ \psi_k \end{bmatrix} [\varphi_s^T \ \psi_s^T]\} = \begin{bmatrix} Q & S \\ S^T & R \end{bmatrix} \delta_{ks}.$$

Nonlinear dynamic output feedback controllers will be considered

$$\begin{cases} z_{k+1} = h(z_k, y_k, d_k) \\ u_k = s(z_k, y_k, d_k) \end{cases} \tag{5.2}$$

with $h(.)$, $s(.)$ continuous nonlinear mappings, $z_k \in \mathbb{R}^{n_z}$ the state of the controller and $d_k \in \mathbb{R}^l$ the reference input.

Nonlinear system identification using neural state space models has been discussed in Chapter 3. Suppose one of the following neural state space models has been identified from I/O measurements on the plant.

**Definition 5.1 [Neural state space models].** The following models $\mathcal{M}_i$ $(i \in \{0,..,3\})$ are called neural state space models

$$\mathcal{M}_0(\theta): \left\{ \begin{array}{rcl} \hat{x}_{k+1} & = & A\hat{x}_k + Bu_k + K\epsilon_k \\ y_k & = & C\hat{x}_k + Du_k + \epsilon_k \end{array} \right.$$

$$\mathcal{M}_1(\theta): \left\{ \begin{array}{rcl} \hat{x}_{k+1} & = & W_{AB}\tanh(V_A\hat{x}_k + V_B u_k + \beta_{AB}) + K\epsilon_k \\ y_k & = & C\hat{x}_k + Du_k + \epsilon_k \end{array} \right.$$

$$\mathcal{M}_2(\theta): \left\{ \begin{array}{rcl} \hat{x}_{k+1} & = & W_{AB}\tanh(V_A\hat{x}_k + V_B u_k + \beta_{AB}) + K\epsilon_k \\ y_k & = & W_{CD}\tanh(V_C\hat{x}_k + V_D u_k + \beta_{CD}) + \epsilon_k \end{array} \right.$$

$$\mathcal{M}_3(\theta): \left\{ \begin{array}{rcl} \hat{x}_{k+1} & = & W_{AB}\tanh(V_{AB}\tanh(V_A\hat{x}_k + V_B u_k + \beta_{ABv}) + \beta_{ABw}) + K\epsilon_k \\ y_k & = & W_{CD}\tanh(V_{CD}\tanh(V_C\hat{x}_k + V_D u_k + \beta_{CDv}) + \beta_{CDw}) + \epsilon_k. \end{array} \right. \tag{5.3}$$

**Remark.** Here tanh(.) has to be applied elementwise. The linear model in innovations form $\mathcal{M}_0$ is the *Kalman filter* (see e.g. Ljung (1987)(1979)) and can be interpreted as a specific neural state space model having a parametrization without hidden layers and a linear activation function for the output neurons. In the deterministic case we have zero Kalman gain $K$. The reason why the linear model $\mathcal{M}_0$ is included as one of the neural state space models is to make a comparison between $\text{NL}_q$ theory and linear control theory in the sequel.

We consider nonlinear dynamic output feedback controllers that are parametrized in a similar way as the neural state space models and are called neural state space controllers.

**Definition 5.2 [Neural state space controllers].** The following controllers $\mathcal{C}_i$ $(i \in \{0,..,5\})$ are called neural state space controllers

$$\mathcal{C}_0(\theta_c): \left\{ \begin{array}{rcl} z_{k+1} & = & Ez_k + Fy_k + F_2 d_k \\ u_k & = & Gz_k + Hy_k + H_2 d_k \end{array} \right.$$

$$\mathcal{C}_1(\theta_c): \left\{ \begin{array}{rcl} z_{k+1} & = & Ez_k + Fy_k + F_2 d_k \\ u_k & = & \tanh(Gz_k + Hy_k + H_2 d_k) \end{array} \right. \tag{5.4}$$

$$\mathcal{C}_2(\theta_c): \left\{ \begin{array}{rcl} z_{k+1} & = & W_{EF}\tanh(V_E z_k + V_F y_k + V_{F_2} d_k + \beta_{EF}) \\ u_k & = & W_{GH}\tanh(V_G z_k + V_H y_k + V_{H_2} d_k + \beta_{GH}) \end{array} \right.$$

$$\mathcal{C}_3(\theta_c): \begin{cases} z_{k+1} &= W_{EF} \tanh(V_E z_k + V_F y_k + V_{F_2} d_k + \beta_{EF}) \\ u_k &= \tanh(W_{GH} \tanh(V_G z_k + V_H y_k + V_{H_2} d_k + \beta_{GH})) \end{cases}$$

$$\mathcal{C}_4(\theta_c): \begin{cases} z_{k+1} &= W_{EF} \tanh(V_{EF} \tanh(V_E z_k + V_F y_k + V_{F_2} d_k + \beta_{EF_v}) + \beta_{EF_w}) \\ u_k &= W_{GH} \tanh(V_{GH} \tanh(V_G z_k + V_H y_k + V_{H_2} d_k + \beta_{GH_v}) + \beta_{GH_w}) \end{cases}$$

$$\mathcal{C}_5(\theta_c): \begin{cases} z_{k+1} &= W_{EF} \tanh(V_{EF} \tanh(V_E z_k + V_F y_k + V_{F_2} d_k + \beta_{EF_v}) + \beta_{EF_w}) \\ u_k &= \tanh(W_{GH} \tanh(V_{GH} \tanh(V_G z_k + V_H y_k + V_{H_2} d_k + \beta_{GH_v}) + \beta_{GI} \end{cases}$$

**Remarks:** The controller $\mathcal{C}_1$, $\mathcal{C}_3$, $\mathcal{C}_5$ correspond respectively to the controller $\mathcal{C}_0$, $\mathcal{C}_2$, $\mathcal{C}_4$ but with saturated output $u_k$. $\mathcal{C}_0$ is a classical linear dynamic output feedback controller (e.g. Boyd & Barratt (1991), Maciejowski (1989)).

Like in modern control theory we will consider the standard plant configuration of Figure 5.3 (see Boyd & Barratt (1991)) instead of the classical control scheme. This standard plant is a reorganized scheme with a so-called exogenous input $w_k$ (consisting of the reference input and disturbance signals), regulated output $e_k$ (consisting of $d_k - \hat{y}_k$ and possibly also other variables of interest), sensed output $y_k$ and actuator input $u_k$. An *augmented plant* $\mathcal{S}_i$ is considered with inputs $w_k$, $u_k$ and output $e_k$, $y_k$.

**Definition 5.3 [Neural state space control problem $\Xi_j^i$].** A neural control problem $\Xi_j^i$ is the control problem related to the control scheme of Figure 5.1 with a neural state space controller $\mathcal{C}_j$ ($j \in \{0, 1, ..., 5\}$) and a neural state space model $\mathcal{M}_i$ ($i \in \{0, 1, 2, 3\}$).

The family of problems $\Xi_j^i$ is given in Table 5.1.

| | $\mathcal{C}_0$ | $\mathcal{C}_1$ | $\mathcal{C}_2$ | $\mathcal{C}_3$ | $\mathcal{C}_4$ | $\mathcal{C}_5$ |
|---|---|---|---|---|---|---|
| $\mathcal{M}_0$ | $\Xi_0^0$ | $\Xi_1^0$ | - | - | - | - |
| $\mathcal{M}_1$ | $\Xi_0^1$ | $\Xi_1^1$ | $\Xi_2^1$ | $\Xi_3^1$ | - | - |
| $\mathcal{M}_2$ | $\Xi_0^2$ | $\Xi_1^2$ | $\Xi_2^2$ | $\Xi_3^2$ | - | - |
| $\mathcal{M}_3$ | $\Xi_0^3$ | $\Xi_1^3$ | $\Xi_2^3$ | $\Xi_3^3$ | $\Xi_4^3$ | $\Xi_5^3$ |

Table 5.1: *This table shows the several model-based neural control strategies $\Xi_j^i$ that are considered within $NL_q$ theory. The rows represent possible models in increasing level of complexity (hidden layers). The columns represent the several possible neural controllers in increasing level of complexity.*

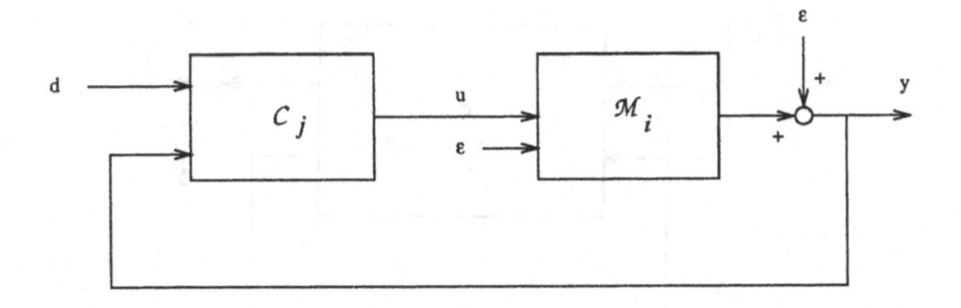


Figure 5.1: *Control scheme with a neural state space model (emulator) $\mathcal{M}_i$ and a neural state space controller $\mathcal{C}_j$. The model $\mathcal{M}_i$ is described by means of a nonlinear state space model of the form $\hat{x}_{k+1} = f(\hat{x}_k, u_k, \epsilon_k; \theta_m)$, $y_k = g(\hat{x}_k, u_k, \epsilon_k; \theta_m)$ and $\mathcal{C}_j$ is a dynamic output feedback controller with nonlinear state space model $z_{k+1} = h(z_k, y_k, d_k; \theta_c)$, $u_k = s(z_k, y_k, d_k; \theta_c)$. The functions $f(.)$, $g(.)$, $h(.)$, $s(.)$ are parametrized by multilayer feedforward neural networks, with interconnection weights $\theta_m$ for $f(.)$, $g(.)$ and $\theta_c$ for $h(.)$, $s(.)$ respectively. Several parametrizations are considered, resulting in a certain neural control problem $\Xi_j^i$ related to a model $\mathcal{M}_i$ and a controller $\mathcal{C}_j$. The signal $y_k$ represents the output of the plant, $d_k$ the reference input, $u_k$ the control signal and $\epsilon_k$ a white noise innovations signal.*

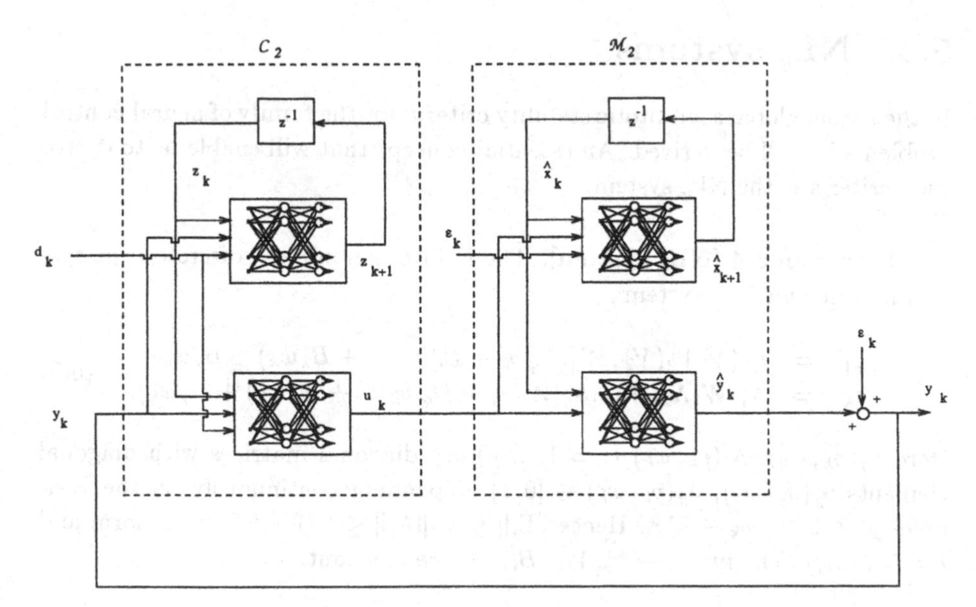


Figure 5.2: *Specific case $\Xi_2^2$ of Figure 5.1.*

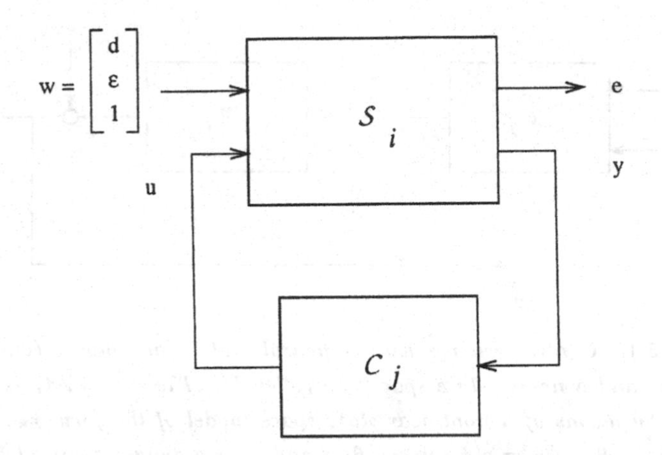


Figure 5.3: *Standard plant representation according to modern control theory with $e_k$ the regulated outputs, $y_k$ the sensed outputs, $u_k$ the actuator inputs and $w_k = [d_k; \epsilon_k; 1]$ the exogenous input consisting of the reference inputs and disturbances. The constant input 1 is due to the bias term of the neural network architectures. $S_i$ is called the augmented plant model.*

## 5.2 $NL_q$ systems

In the sequel global asymptotic stability criteria for the family of neural control problems $\Xi^i_j$ will be derived. An essential concept that will enable us to derive such criteria is the $NL_q$ system.

**Definition 5.4 [$NL_q$ system].** The following nonlinear discrete time system is called an $NL_q$ system

$$\left\{ \begin{array}{rcl} p_{k+1} & = & \Gamma_1( V_1 \, \Gamma_2( V_2 \ldots \Gamma_q( V_q \, p_k + B_q w_k) \ldots + B_2 w_k) + B_1 w_k) \\ e_k & = & \Lambda_1(W_1 \Lambda_2(W_2 \ldots \Lambda_q(W_q p_k + D_q w_k) \ldots + D_2 w_k) + D_1 w_k). \end{array} \right. \tag{5.5}$$

Here $\Gamma_i(p_k, w_k)$, $\Lambda_i(p_k, w_k)$ $(i = 1, ..., q)$ are diagonal matrices with diagonal elements $\gamma_j(p_k, w_k)$, $\lambda_j(p_k, w_k) \in [0, 1]$, depending continuously on the variables $p_k \in \mathbb{R}^{n_p}$, $w_k \in \mathbb{R}^{n_w}$. Hence $||\Gamma_i|| \leq 1$, $||\Lambda_i|| \leq 1$ (for 1,2 or $\infty$ norm and $i = 1, 2, ...q$). The matrices $V_i$, $W_i$, $B_i$, $D_i$ are constant.

**Remark.** The term '$NL_q$' system refers to the alternating sequence of nonlinear and linear operators acting on the input arguments $p_k, w_k$: the diagonal matrices $\Gamma_i$, $\Lambda_i$ depend on $p_k, w_k$ and refer to 'N' in '$NL_q$', the constant matrices $V_i$, $W_i$, $B_i$, $D_i$ refer to 'L' in '$NL_q$'. The index $q$ refers to the number of nonlinear operations in this alternating sequence and is a measure of complexity of the overall system.

Such $NL_q$ systems (or shortly $NL_q$s) will then be considered for the standard plant representation of Figure 5.3. In order to fix the ideas, the cases $q = 1, 2$ are of the form

$$\begin{aligned} q=1: &\quad \begin{cases} p_{k+1} = \Gamma_1(V_1 p_k + B_1 w_k) \\ e_k = \Lambda_1(W_1 p_k + D_1 w_k) \end{cases} \\ q=2: &\quad \begin{cases} p_{k+1} = \Gamma_1(V_1 \Gamma_2(V_2 p_k + B_2 w_k) + B_1 w_k) \\ e_k = \Lambda_1(W_1 \Lambda_2(W_2 p_k + D_2 w_k) + D_1 w_k). \end{cases} \end{aligned} \tag{5.6}$$

The $NL_q$ system is related to a recurrent network of the form

$$\begin{cases} p_{k+1} = \sigma(V_1\, \sigma(V_2 \ldots \sigma(V_q\, p_k + B_q w_k) \ldots + B_2 w_k) + B_1 w_k) \\ e_k = \sigma(W_1\, \sigma(W_2 \ldots \sigma(W_q p_k + D_q w_k) \ldots + D_2 w_k) + D_1 w_k) \end{cases} \tag{5.7}$$

where $\sigma(.)$ is a static nonlinear operator that satisfies the sector condition [0, 1]. Special cases for $\sigma(.)$ are e.g. tanh(.) and sat(.) (defined as $\text{sat}(x) = x$ if $|x| \leq 1$ and $\text{sat}(x) = \text{sign}(x)$ if $|x| \geq 1$). The fact that this $NL_q$ is related to nonlinear operators that satisfy a sector condition can be understood as follows. Consider a system related to the $NL_1$

$$x_{k+1} = \sigma(W x_k). \tag{5.8}$$

This can be written as

$$x_{k+1} = \Gamma(x_k) W x_k$$

with $\Gamma = \text{diag}\{\gamma_i\}$ and $\gamma_i = \sigma(w_i^T x_k)/(w_i^T x_k)$. This is easily obtained by using an elementwise notation and based on the fact that tanh(.) is a diagonal nonlinearity. Eqn (5.8) becomes

$$\begin{aligned} x^i &:= \sigma(\textstyle\sum_j w^i_j x^j) \\ &:= \frac{\sigma(\sum_j w^i_j x^j)}{\sum_j w^i_j x^j} \textstyle\sum_j w^i_j x^j \\ &:= \gamma^i_i \textstyle\sum_j w^i_j x^j. \end{aligned}$$

The time index is omitted here because of the assignment operator ':='. The notation $\gamma^i_i$ means that this corresponds to the diagonal matrix $\Gamma(x_k)$. In case $w_i^T x_k = 0$ de l' Hospital's rule or a Taylor expansion of $\sigma(.)$ leads to $\gamma_i = 1$.

For an additional layer

$$x_{k+1} = \sigma(V\sigma(Wx_k)), \tag{5.9}$$

one obtains

$$x_{k+1} = \Gamma_1(x_k)V\Gamma_2(x_k)Wx_k$$

where $\Gamma_1 = \mathrm{diag}\{\gamma_{1_i}\}$ with $\gamma_{1_i} = \sigma(v_i^T\sigma(Wx_k))/(v_i^T\sigma(Wx_k))$ and $\Gamma_2 = \mathrm{diag}\{\gamma_{2_i}\}$ with $\gamma_{2_i} = \sigma(w_i^T x_k)/(w_i^T x_k)$. Indeed from an elementwise notation one has

$$\begin{aligned} x^i &:= \sigma(\textstyle\sum_j v_j^i \sigma(\sum_l w_l^j x^l) \\ &:= \sigma(\textstyle\sum_j v_j^i \gamma_{2_j}^j \sum_l w_l^j x^l) \\ &:= \gamma_{1_i}^i \textstyle\sum_j v_j^i \gamma_{2_j}^j \sum_l w_l^j x^l. \end{aligned}$$

The latter has become clear in Suykens *et al.* (1993e) by using the Einstein summation convention.

The aim is now to bring neural state space models and neural control systems in the standard NL$_q$ form.

**Lemma 5.1 [$\mathcal{M}_i$'s are NL$_q$s].** Neural state space models $\mathcal{M}_i$ ($i \in \{0, 1, ..., 3\}$) can be written as NL$_q$ systems with $q = 1$ if $i = 0$ and $q = 2$ if $i = 1, 2$ and finally $q = 3$ if $i = 3$. The input vector of the NL$_q$ is $w_k = [u_k; \epsilon_k; 1]$.

*Proof:* See Appendix C. By taking an elementwise notation, multiplying and dividing by the arguments of the activation functions, the alternating sequence of nonlinear and linear operators in (5.5) is obtained. The elements $\gamma_i$, $\lambda_i$ belong to [0, 1] because the activation function tanh is a nonlinearity belonging to sector [0,1].

□

**Remark.** The constant input 1 in $w_k$ is due to the bias vector of the neural network. It is well known that the bias vector can be treated formally as part of the interconnection matrix by introducing an additional constant input to the network.

**Lemma 5.2 [$\Xi_j^i$'s are NL$_q$s].** All members of the family of neural control problems $\Xi_j^i$ can be written as NL$_q$ systems.

*Proof:* An exhaustive list is presented in Suykens *et al.* (1994f). Some examples will be discussed below. A trick of state augmentation is used in order to achieve the result.

□

The Lemma is illustrated here with the cases $\Xi_1^0$ and $\Xi_2^2$.

**Example 5.1.** Given the Kalman filter ($\mathcal{M}_0$) and a linear dynamic output feedback controller with saturation ($\mathcal{C}_1$)

$$\mathcal{M}_0 : \begin{cases} \hat{x}_{k+1} &= A\hat{x}_k + Bu_k + K\epsilon_k \\ y_k &= C\hat{x}_k + Du_k + \epsilon_k \end{cases}$$

$$\mathcal{C}_1 : \begin{cases} z_{k+1} &= Ez_k + Fy_k + F_2 d_k \\ u_k &= \tanh(Gz_k) \end{cases}$$

the state equation for the closed loop system is

$$\begin{cases} \hat{x}_{k+1} &= A\hat{x}_k + B\tanh(Gz_k) + K\epsilon_k \\ z_{k+1} &= FC\hat{x}_k + Ez_k + FD\tanh(Gz_k) + F\epsilon_k + F_2 d_k. \end{cases}$$

A new state variable is introduced

$$\xi_k = \tanh(Gz_k).$$

This leads to a state augmentation of the closed loop system

$$\begin{cases} \hat{x}_{k+1} &= A\hat{x}_k + B\xi_k + K\epsilon_k \\ z_{k+1} &= Ez_k + FC\hat{x}_k + FD\xi_k + F\epsilon_k + F_2 d_k \\ \xi_{k+1} &= \tanh(GEz_k + GFC\hat{x}_k + GFD\xi_k + GF\epsilon_k + GF_2 d_k) \end{cases}$$

and can be written as an $\text{NL}_1$ system

$$p_{k+1} = \Gamma_1(V_1 p_k + B_1 w_k)$$

by taking $p_k = [\hat{x}_k; z_k; \xi_k]$, $w_k = [d_k; \epsilon_k; 1]$ and matrices

$$V_1 = \begin{bmatrix} A & 0 & B \\ FC & E & FD \\ GFC & GE & GFD \end{bmatrix}, B_1 = \begin{bmatrix} 0 & K & 0 \\ F_2 & F & 0 \\ GF_2 & GF & 0 \end{bmatrix}$$

and $\Gamma_1 = \text{diag}\{I, I, \Gamma_G(\xi_k, z_k, \hat{x}_k, d_k, \epsilon_k)\}$, where $\Gamma_G$ is a diagonal matrix with elements $\gamma_{G_i} \in [0, 1]$ which follows from Lemma 5.1.

**Example 5.2.** Given the model $\mathcal{M}_2$ and the controller $\mathcal{C}_2$

$$\mathcal{M}_2 : \begin{cases} \hat{x}_{k+1} &= W_{AB}\tanh(V_A\hat{x}_k + V_B u_k + \beta_{AB}) + K\epsilon_k \\ y_k &= W_C\tanh(V_C\hat{x}_k) + \epsilon_k \end{cases}$$

$$\mathcal{C}_2 : \begin{cases} z_{k+1} &= W_{EF}\tanh(V_E z_k + V_F y_k + V_{F_2} d_k + \beta_{EF}) \\ u_k &= W_G\tanh(V_G z_k) \end{cases}$$

the state equation for the closed loop system is

$$\begin{cases} \hat{x}_{k+1} &= W_{AB}\tanh(V_A\hat{x}_k + V_B W_G \tanh(V_G z_k) + \beta_{AB}) + K\epsilon_k \\ z_{k+1} &= W_{EF}\tanh(V_E z_k + V_F W_C \tanh(V_C\hat{x}_k) + V_F\epsilon_k + V_{F_2} d_k + \beta_{EF}). \end{cases}$$

State augmentation is done by defining $\xi_k = \tanh(V_C\hat{x}_k)$ and $\eta_k = \tanh(V_G z_k)$, which gives

$$\begin{cases} \hat{x}_{k+1} &= W_{AB}\tanh(V_A\hat{x}_k + V_B W_G \eta_k + \beta_{AB}) + K\epsilon_k \\ z_{k+1} &= W_{EF}\tanh(V_E z_k + V_F W_C \xi_k + V_F\epsilon_k + V_{F_2} d_k + \beta_{EF}) \\ \xi_{k+1} &= \tanh(V_C W_{AB}\tanh(V_A\hat{x}_k + V_B W_G\eta_k + \beta_{AB}) + V_C K\epsilon_k) \\ \eta_{k+1} &= \tanh(V_G W_{EF}\tanh(V_E z_k + V_F W_{CD}\xi_k + V_F\epsilon_k + V_{F_2} d_k + \beta_{EF})) \end{cases}$$

and can be written as an NL$_2$ system

$$p_{k+1} = \Gamma_1(V_1\Gamma_2(V_2 p_k + B_2 w_k) + B_1 w_k)$$

by taking $p_k = [\hat{x}_k; z_k; \xi_k; \eta_k]$, $w_k = [d_k; \epsilon_k; 1]$ and

$$V_1 = \begin{bmatrix} W_{AB} & & & \\ & W_{EF} & & \\ & & V_C W_{AB} & \\ & & & V_G W_{EF} \end{bmatrix}, V_2 = \begin{bmatrix} V_A & 0 & 0 & V_B W_G \\ 0 & V_E & V_F W_C & 0 \\ V_A & 0 & 0 & V_B W_G \\ 0 & V_E & V_F W_C & 0 \end{bmatrix}$$

$$B_2 = \begin{bmatrix} 0 & 0 & \beta_{AB} \\ V_{F_2} & V_F & \beta_{EF} \\ 0 & 0 & \beta_{AB} \\ V_{F_2} & V_F & \beta_{EF} \end{bmatrix}, B_1 = \begin{bmatrix} 0 & K & 0 \\ 0 & 0 & 0 \\ 0 & V_C K & 0 \\ 0 & 0 & 0 \end{bmatrix}$$

In Table 5.2 the resulting $q$ value for the NL$_q$ related to a given $\Xi_j^i$ problem is shown, together with the number of additional state variables that is needed to obtain the NL$_q$ form.

| | $\mathcal{C}_0$ | $\mathcal{C}_1$ | $\mathcal{C}_2$ | $\mathcal{C}_3$ | $\mathcal{C}_4$ | $\mathcal{C}_5$ |
|---|---|---|---|---|---|---|
| $\mathcal{M}_0$ | 1 | 1 | - | - | - | - |
| $\mathcal{M}_1$ | 2 | 2 | 2 | 3 | - | - |
| $\mathcal{M}_2$ | 2 | 2 | 2 | 3 | - | - |
| $\mathcal{M}_3$ | 4 | 4 | 4 | 4 | 4 | 5 |

| | $\mathcal{C}_0$ | $\mathcal{C}_1$ | $\mathcal{C}_2$ | $\mathcal{C}_3$ | $\mathcal{C}_4$ | $\mathcal{C}_5$ |
|---|---|---|---|---|---|---|
| $\mathcal{M}_0$ | 0 | 1 | - | - | - | - |
| $\mathcal{M}_1$ | 0 | 1 | 1 | 1 | - | - |
| $\mathcal{M}_2$ | 1 | 2 | 2 | 2 | - | - |
| $\mathcal{M}_3$ | 1 | 2 | 2 | 2 | 2 | 2 |

Table 5.2: *Neural state space control problems $\Xi_j^i$: (Left) corresponding $q$ value for the NL$_q$ related to the problem $\Xi_j^i$; (Right) the number of new state variables to be introduced in order to obtain the NL$_q$ form for a given $\Xi_j^i$.*

## 5.3 Global asymptotic stability criteria for $NL_q$s

We investigate here the autonomous case of $NL_q$ systems for the standard plant configuration. Sufficient conditions for global asymptotic stability will be derived. Furthermore we explain how the discrete time Lur'e problem from classical nonlinear control theory fits into the present $NL_q$ framework.

### 5.3.1 Stability criteria

The following Theorem holds for autonomous $NL_q$s (zero external input, i.e. $w_k = 0$ in (5.5)):

**Theorem 5.1 [Diagonal scaling].** A sufficient condition for global asymptotic stability of the autonomous $NL_q$ system

$$p_{k+1} = (\prod_{i=1}^{q} \Gamma_i(p_k) V_i)\, p_k \tag{5.10}$$

is to find diagonal matrices $D_i$ such that

$$\|D_{tot} V_{tot} D_{tot}^{-1}\|_2^q = \beta_D < 1 \tag{5.11}$$

where the matrices $V_{tot}$ and $D_{tot}$ are given by:

–

$$V_{tot} = \begin{bmatrix} 0 & V_2 & & & 0 \\ & 0 & V_3 & & \\ & & \ddots & & \\ & & & 0 & V_q \\ V_1 & & & & 0 \end{bmatrix}, V_i \in \mathbb{R}^{n_{h_i} \times n_{h_{i+1}}}, n_{h_1} = n_{h_{q+1}} = n_p$$

– $D_{tot} = \text{diag}\{D_2, D_3, ..., D_q, D_1\}$

– $D_i \in \mathbb{R}^{n_{h_i} \times n_{h_i}}$ diagonal matrices with nonzero diagonal elements.

*Proof:* See Appendix C. The Theorem is based on the Lyapunov function $V = \|D_1 p\|_2$ which is radially unbounded. The proof makes use of the properties of induced norms.

□

**Remarks.**

- The equilibrium point $p = 0$ in (5.10) is unique because the Lyapunov function is radially unbounded.

- The minimum over $D_{tot}$ yields a maximal contraction rate of the flow for the $NL_q$ system and shows how such a feasible point $D_{tot}$ is found in practice as

$$\min_{D_{tot}} \beta_D. \tag{5.12}$$

- Given a constant matrix $V_{tot}$ the criterion (5.11) in the unknown diagonal matrix $\log(D_{tot})$ is a *convex optimization problem.* This implies that the problem has only one minimum, which is the global one. Moreover this minimum can be found in polynomial time. One also meets criteria of the form (5.11) in the field of modern robust control theory (e.g. Boyd & Yang (1989), Doyle (1982), Packard & Doyle (1993), Kaszkurewicz & Bhaya (1993)). This already suggests there are links between that field and the neural state space model framework that is outlined here. The precise links will be pointed out in Section 5.5.

From robust control theory (see e.g. Packard & Doyle (1993)) it is well-known that conditions of diagonal scaling might be too conservative. Hence we will try to find sharper conditions by allowing full matrices $P_i$ instead of diagonal matrices $D_i$. In the following Theorem diagonality is relaxed to diagonal dominance of the matrices $P_i^T P_i$. In order to prove the next Theorem, we need first the following definition.

**Definition 5.5 [Level of diagonal dominance].** A matrix $Q \in \mathbb{R}^{n \times n}$ is called diagonally dominant of level $\delta_Q \geq 1$ if the following property holds:

$$q_{ii} > \delta_Q . \sum_{j=1\,(j \neq i)}^{n} |q_{ij}| \quad , \quad \forall i = 1, ..., n. \tag{5.13}$$

**Remark.** This definition is consistent with the original definition of diagonal dominance (see e.g. Liu & Michel (1992)), which corresponds to the special case $\delta_Q = 1$.

**Theorem 5.2 [Diagonal dominance].** A sufficient condition for global asymptotic stability of the autonomous $NL_q$ system (5.10) is to find matrices $P_i$, $N_i$ such that

$$c_\alpha \, \beta_P < 1 \tag{5.14}$$

with $c_\alpha = \prod_{i=1}^{q} (1 + \alpha_i)^{1/2}$ and $\beta_P = \|P_{tot} V_{tot} P_{tot}^{-1}\|_2^q$, where the matrix $V_{tot}$ is given by (5.11) and $P_{tot}$ by

- $P_{tot} = \text{blockdiag}\{P_2, P_3, ..., P_q, P_1\}$, $P_i \in \mathbb{R}^{n_{h_i} \times n_{h_i}}$ of full rank.

- $Q_i = P_i^T P_i N_i$ are diagonally dominant matrices with $\delta_{Q_i} = (1+\alpha_i)/\alpha_i \geq 1 \quad (\alpha_i \geq 0)$ and $N_i$ diagonal with positive diagonal elements.

*Proof:* See Appendix C. The proof is based on the Lyapunov function $V(p) = \|P_1 p\|_2$ which is radially unbounded and makes use of the properties of induced norms. The levels of diagonal dominance $\delta_{Q_i}$ are derived from Gerschgorin's Theorem (see Wilkinson (1965)).

□

The following Lemma gives an equivalent expression for diagonal dominance of a matrix in terms of its diagonal and its off-diagonal part:

**Lemma 5.3.** Given a matrix $Q \in \mathbb{R}^{n \times n}$, the condition of diagonal dominance of level $\delta_Q$

$$q_{ii} > \delta_Q . \sum_{j=1\,(j \neq i)}^{n} |q_{ij}| \quad , \quad \forall i = 1, ..., n$$

is equivalent to the condition

$$\|X_Q\|_\infty < 1/\delta_Q \tag{5.15}$$

with $X_Q = D_Q^{-1} H_Q$, where $Q = D_Q + H_Q$ and $D_Q$ is the diagonal and $H_Q$ the off-diagonal part of $Q$.

*Proof:* See Appendix C. It follows immediately from the definition of $\|X_Q\|_\infty$.

□

**Remarks.**

- Again the equilibrium point $p = 0$ is unique because the Lyapunov function is radially unbounded.
- The solution to the optimization problem

$$\min_{P_i, N_i} c_\alpha \beta_P \tag{5.16}$$

  yields a maximal contraction rate of the flow of the autonomous $\mathrm{NL}_q$ system with respect to the condition (5.14).
- Theorem 5.1 is a special case of Theorem 5.2 corresponding to $\alpha_i = 0$ (or $\delta_{Q_i} \to \infty$), making the matrices $P_i$ diagonal.

- The problem (5.14) can be interpreted as follows: find matrices $P_i, N_i$ such that

$$||P_{tot} V_{tot} P_{tot}^{-1}||_2^q < 1/c_\alpha$$

and

$$||X_{Q_i}||_\infty < 1/\delta_{Q_i}.$$

The upper bounds are plotted as a function of $\alpha_i$ for $q = 1$ in Figure 5.4. There exists a trade-off between $\delta_Q$ and $c_\alpha$: the lower the level of diagonal dominance on the $Q_i$'s is, the stronger the condition on $||P_{tot} V_{tot} P_{tot}^{-1}||_2$ becomes. Low diagonal dominance without a large correction factor $c_\alpha$ is typically infeasible.

- Elimination of $\alpha_i$ in (5.14), yields the criterion

$$\prod_{i=1}^{q} \left(\frac{\delta_{Q_i}}{\delta_{Q_i} - 1}\right)^{1/2} ||P_{tot} V_{tot} P_{tot}^{-1}||_2^q < 1.$$

- For all values $l \in \{1, 2, ..., q\}$ for which $\Gamma_l = I$ there is no condition of diagonal dominance on $Q_l$, but only on the remaining $Q_i$'s ($i \neq l$). This follows immediately from the proof of the Theorem. Examples of this are the neural state space control problems $\Xi_0^1$ and $\Xi_1^1$ for which $\Gamma_1 = I$. Another example is the linear system

$$p_{k+1} = A p_k,$$

for which a sufficient condition for global asymptotic stability becomes

$$\min_{P \in \mathbb{R}^{n \times n}} ||PAP^{-1}||_2 < 1 \qquad (5.17)$$

because $\Gamma_1 = I$ and by taking the Lyapunov function $V(p) = ||Pp||_2$. It is also well known that this corresponds to

$$\rho(A) < 1$$

where $\rho(A)$ denotes the spectral radius of $A$ (see Boyd *et al.* (1994)). In this case the condition (5.17) is also necessary.

- In the case of a sat(.) activation function it is proven in Liu & Michel (1992) for $q = 1$ that a necessary and sufficient condition for

$$\text{sat}(p)^T Q \, \text{sat}(p) < p^T Q \, p$$

to hold $\forall p$ is $\delta_Q = 1$, taking $N_1 = I$. Hence for that specific activation function (5.14) becomes for $q = 1$

$$\|PVP^{-1}\|_2 < 1 \tag{5.18}$$

subject to

$$\|X_Q\|_\infty \leq 1.$$

which follows from the Proof of Theorem 5.2. Hence $c_\alpha = 1$ for this particular activation function. This problem was investigated in the context of the stability of digital filters with overflow characteristic in Liu & Michel (1992).

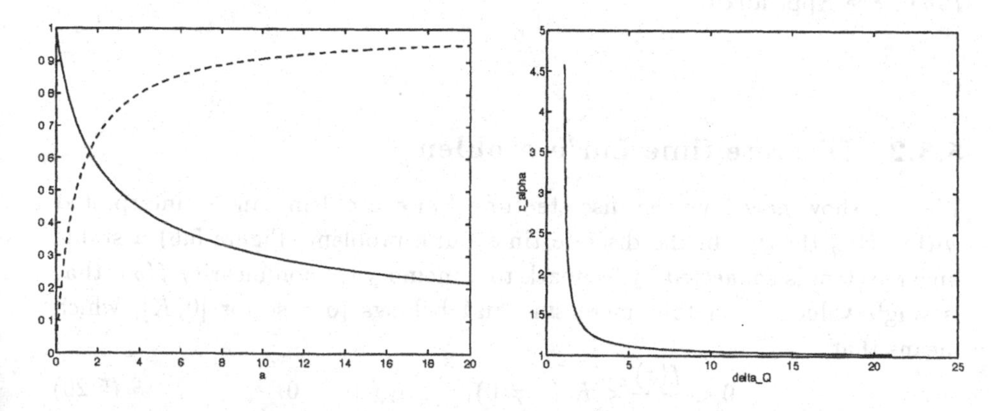


Figure 5.4: *Interpretation of Theorem 5.2 (diagonal dominance case) for global asymptotic stability of autonomous $NL_q$s with $q = 1$: (Left) Upper bounds $1/c_\alpha$ (full line) and $1/\delta_Q$ (dashed line) as a function of $\alpha$. (Right) Trade-off curve between $c_\alpha$ and $\delta_Q$. The curve shows that a high correction factor $c_\alpha$ is needed if the level of diagonal dominance of $Q = P^T P$ becomes low. Large diagonal dominance of $Q$ implies a small correction factor $c_\alpha$. The limiting case $\delta_Q \to \infty$ corresponds to diagonal scaling.*

The following Theorem allows also full instead of diagonal matrices $P_i$, but a correction factor is expressed in terms of the condition numbers of the matrices $P_i$.

**Theorem 5.3 [Condition number factor].** A sufficient condition for global asymptotic stability of the autonomous $NL_q$ is to find matrices $P_i$ such that

$$\prod_{i=1}^{q} \kappa(P_i) \, ||P_{tot} V_{tot} P_{tot}^{-1}||_2^q < 1 \tag{5.19}$$

where $V_{tot}$ is given by (5.11) and $P_{tot} = \text{blockdiag}\{P_2, P_3, ..., P_q, P_1\}$ with $P_i \in \mathbb{R}^{n_{h_i} \times n_{h_i}}$ full rank matrices. The condition numbers $\kappa(P_i)$ are by definition equal to $||P_i||_2 ||P_i^{-1}||_2$.

*Proof:* See Appendix C.

□

## 5.3.2 Discrete time Lur'e problem

We will show now how the discrete time Lur'e problem can be interpreted within $NL_q$ theory. In the discrete time Lur'e problem (Figure 5.5) a stable linear system is connected by feedback to a memoryless nonlinearity $f(\sigma)$, that is single valued, monotone increasing and belongs to a sector $[0, K]$, which means that

$$0 < \frac{f(\sigma)}{\sigma} < K \;(\sigma \neq 0), \qquad f(0) = 0. \tag{5.20}$$

An investigation on the stability of such systems started around 1960 e.g. in the work of Tsypkin, Szegö, Pearson and Gibson, Jury and Lee. For a survey on this subject see, e.g., Brockett (1966) and Vidal (1969). The following Theorem by Tsypkin can be seen as the closest analog for discrete time systems to Popov's result for the corresponding continuous time case.

**Theorem 5.4 [Tsypkin].** Let $p(z)$ and $q(z)$ denote polynomials without common factors and let

$$\begin{cases} x_{k+1} &= Ax_k + bu_k \\ y_k &= c^T x_k \end{cases}$$

be a minimal representation of the transfer function of the linear part $G(z) = q(z)/p(z)$. If the zeros of $p(z)$ lie inside the disk $|z| = 1$, then the equilibrium point $x = 0$ is globally asymptotically stable, if $0 \leq f(y)/y \leq K$ and

$$Re[G(z)] + \frac{1}{K} > 0 \ \forall z : |z| = 1$$

*Proof:* See e.g. Jury (1974).

□

Within $NL_q$ theory the Lur'e problem can be formulated as follows:

**Theorem 5.5 [Lur'e problem as NL$_1$ system].** Given the discrete time Lur'e problem for a multivariable system with $m$ inputs and $m$ outputs, this can be formulated within the neural state space model framework as

$$\mathcal{M}_0 : \left\{ \begin{array}{rcl} x_{k+1} & = & Ax_k + Bu_k \\ y_k & = & Cx_k \end{array} \right. \qquad \mathcal{C} : \quad u_k = -f(y_k) \tag{5.21}$$

with memoryless, single valued, monotone increasing diagonal nonlinearity $f(\sigma)$, belonging to a sector $[0, K]$ (assume $K \geq 1$). One of the following conditions is sufficient for global asymptotic stability of the system (5.21)

- (Diagonal scaling)

$$K < 1/\beta_D, \qquad \beta_D = \min_D \|DV_{tot}D^{-1}\|_2 \tag{5.22}$$

- (Diagonal dominance)

$$K < 1/(c_\alpha \beta_P), \qquad c_\alpha \beta_P = \min_{P,N} (\delta_Q/(\delta_Q - 1))^{1/2} \|PV_{tot}P^{-1}\|_2 \tag{5.23}$$

- (Condition number factor)

$$K < 1/(c_P \beta_P), \qquad c_P \beta_P = \min_P \kappa(P) \|PV_{tot}P^{-1}\|_2. \tag{5.24}$$

Here $V_{tot}$ is given by

$$V_{tot} = \begin{bmatrix} A & -B \\ CA & -CB \end{bmatrix}$$

and $D$ is a diagonal matrix with nonzero diagonal elements, $Q = P^T PN$ with $P$ a full rank matrix. The level of diagonal dominance of $Q$ is $\delta_Q \geq 1$ and $N$ is a diagonal matrix with positive diagonal elements.

*Proof:* See Appendix C.

□

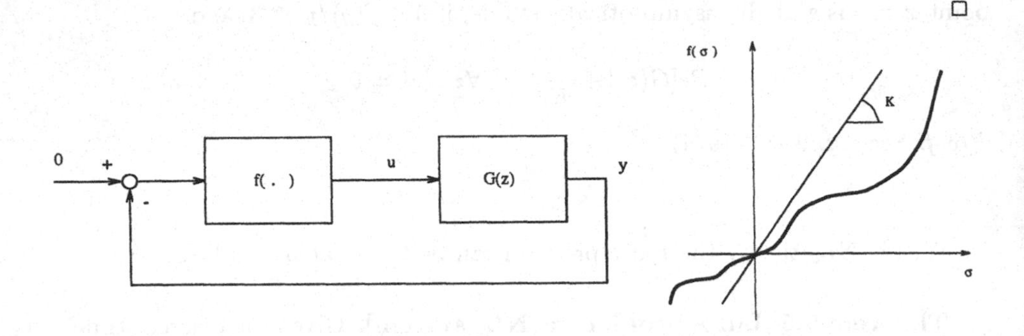


Figure 5.5: *Discrete time Lur'e problem from classical nonlinear control theory, which consists of a stable linear dynamical system, feedback interconnected to a static nonlinearity $f(.)$ belonging to sector $[0, K]$.*

## 5.4 Input/Output properties - $l_2$ theory

In this section input/output properties of the non-autonomous $\mathrm{NL}_q$ system ($w_k \neq 0$ in (5.5)) are studied. First some I/O equivalent representations for $\mathrm{NL}_q$s are formulated. Then an $l_2$ theory is presented. The link between input/output stability and internal stability is clarified through the concept of dissipativity.

### 5.4.1 Equivalent representations for $\mathrm{NL}_q$s

**Lemma 5.4 [Equivalent representations for $\mathrm{NL}_q$s].** I/O equivalent representations for the $\mathrm{NL}_q$ system (5.5) are

$$\left\{ \begin{array}{rcl} p_{k+1} & = & (\prod_{i=1}^{q} \Gamma_{i,e}(p_k, w_k) M_i) \begin{bmatrix} p_k \\ w_k \end{bmatrix} \\ e_k & = & (\prod_{i=1}^{q} \Lambda_{i,e}(p_k, w_k) N_i) \begin{bmatrix} p_k \\ w_k \end{bmatrix} \end{array} \right. \tag{5.25}$$

with

$$\Gamma_{1,e} = \Gamma_1, \quad \Gamma_{i,e} = \begin{bmatrix} \Gamma_i & 0 \\ 0 & I \end{bmatrix}, \quad M_1 = [V_1 \; B_1], \quad M_i = \begin{bmatrix} V_i & B_i \\ 0 & I \end{bmatrix} \; (i = 2, ..., q)$$

$$\Lambda_{1,e} = \Lambda_1, \quad \Lambda_{i,e} = \begin{bmatrix} \Lambda_i & 0 \\ 0 & I \end{bmatrix}, \quad N_1 = [W_1 \; D_1], \quad N_i = \begin{bmatrix} W_i & D_i \\ 0 & I \end{bmatrix} \; (i = 2, ..., q)$$

and

$$\begin{bmatrix} p_{k+1} \\ e_k^{ext} \end{bmatrix} = \left(\prod_{i=1}^{q} \Omega_i(p_k, w_k) R_i\right) \begin{bmatrix} p_k \\ w_k \end{bmatrix} \tag{5.26}$$

with

$$\Omega_i = \begin{bmatrix} \Gamma_{i,e} & 0 & 0 \\ 0 & \Lambda_{i,e} & 0 \\ 0 & 0 & 0 \end{bmatrix}, R_i = \begin{bmatrix} M_i & 0 & 0 \\ 0 & N_i & 0 \\ 0 & 0 & 0 \end{bmatrix}, R_q = \begin{bmatrix} M_q \\ N_q \\ 0 \end{bmatrix} \quad (i = 1, ..., q-1).$$

$e_k^{ext}$ is defined such that its dimension is equal to the dimension of $w_k$ ($e_k^{ext}$ corresponds to $e_k$, augmented with zero elements), in order to make the matrix $\prod_{i=1}^{q} R_i$ square.

*Proof:* straightforward calculation. Remark that $\|\Omega_i\| \leq 1$ because $\|\Gamma_i\| \leq 1$ and $\|\Lambda_i\| \leq 1$.

□

This I/O equivalent form is shown in Figure 5.6. In order to fix the ideas Lemma 5.4 is illustrated for $q = 2$:

$$\begin{cases} p_{k+1} &= \Gamma_1( V_1 \Gamma_2( V_2 p_k + B_2 w_k) + B_1 w_k) \\ e_k &= \Lambda_1 (W_1 \Lambda_2 (W_2 p_k + D_2 w_k) + D_1 w_k) \end{cases}$$

with

$$\begin{cases} p_{k+1} &= \Gamma_1 [V_1 \ B_1] \begin{bmatrix} \Gamma_2 & 0 \\ 0 & I \end{bmatrix} \begin{bmatrix} V_2 & B_2 \\ 0 & I \end{bmatrix} \begin{bmatrix} p_k \\ w_k \end{bmatrix} \\ \\ e_k &= \Lambda_1 [W_1 \ D_1] \begin{bmatrix} \Lambda_2 & 0 \\ 0 & I \end{bmatrix} \begin{bmatrix} W_2 & D_2 \\ 0 & I \end{bmatrix} \begin{bmatrix} p_k \\ w_k \end{bmatrix} \end{cases}$$

and

$$\begin{bmatrix} p_{k+1} \\ e_k^{ext} \end{bmatrix} = \begin{bmatrix} \Gamma_1 & & \\ & \Lambda_1 & \\ & & 0 \end{bmatrix} \begin{bmatrix} V_1 & B_1 & & & \\ & & W_1 & D_1 & \\ & & & & 0 \end{bmatrix} \begin{bmatrix} \Gamma_2 & & & & \\ & I & & & \\ & & \Lambda_2 & & \\ & & & I & \\ & & & & 0 \end{bmatrix} \begin{bmatrix} V_2 & B_2 \\ 0 & I \\ W_2 & D_2 \\ 0 & I \\ 0 & 0 \end{bmatrix} \begin{bmatrix} p_k \\ w_k \end{bmatrix}$$

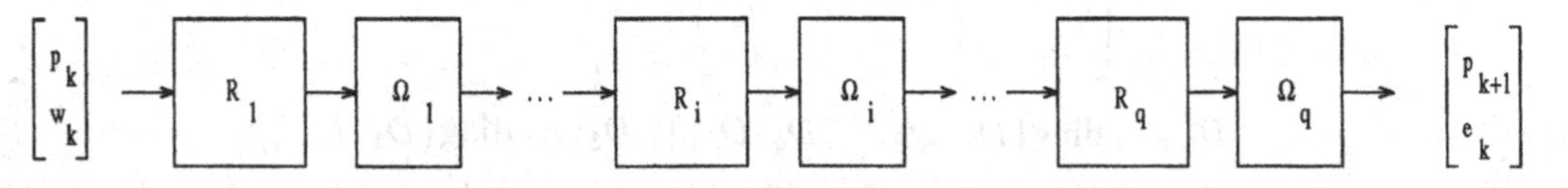


Figure 5.6: *I/O equivalent representation for $NL_q$ system with the typical $q$ 'layer' feature. The matrices $R_i$ are constant and the matrices $\Omega_i(p_k, w_k)$ are diagonal and satisfy the condition $\|\Omega_i\| \leq 1$ for all values of $p_k$, $w_k$. The $NL_q$ system has state $p_k$, input $w_k$ and output $e_k$.*

### 5.4.2 Main Theorems

The following Theorems state sufficient conditions for I/O stability of $NL_q$s.

**Theorem 5.6 [$l_2$ theory - Diagonal scaling].** Given the representation (5.26), if there exist matrices $D_i$ such that

$$\|D_{tot} R_{tot} D_{tot}^{-1}\|_2^q = \beta_D < 1, \tag{5.27}$$

then the following holds:

1. For a finite time horizon $N$:

$$\sum_{k=0}^{N-1} \|e_k\|_2^2 \leq \beta_D^2 \sum_{k=0}^{N-1} \|w_k\|_2^2 + r_0^2 \tag{5.28}$$

where $r_0 = \|D_1 p_0\|_2$.

2. For $N \to \infty$:
There exist constants $c_1, c_2$ such that

$$c_2(1-\beta_D^2)\|p\|_2^2 + \|e\|_2^2 \leq \beta_D^2 \|w\|_2^2 + c_1 \|p_0\|_2^2 \tag{5.29}$$

provided that $\{w_k\}_{k=0}^{\infty} \in l_2$. Hence also

$$\|e\|_2^2 \leq \beta_D^2 \|w\|_2^2 + c_1 \|p_0\|_2^2$$

holds in that case.

Here $l_2^n$ denotes the set of square summable sequences in $\mathbb{C}^n$. The matrices $R_{tot}$ and $D_{tot}$ are given by:

–

$$R_{tot} = \begin{bmatrix} 0 & R_2 & & & 0 \\ & 0 & R_3 & & \\ & & \ddots & & \\ & & & 0 & R_q \\ R_1 & & & & 0 \end{bmatrix}, R_i \in \mathbb{R}^{n_{r_i} \times n_{r_{i+1}}}, n_{r_1} = n_{r_{q+1}} = n_p + n_w$$

– $D_{tot} = \text{diag}\{D_2, D_3, ..., D_q, D_{S_1}\}$, $D_{S_1} = \text{diag}\{D_1, I_{n_w}\}$

– $D_1 \in \mathbb{R}^{n_p \times n_p}$, $D_i \in \mathbb{R}^{n_{r_i} \times n_{r_i}}$ diagonal matrices with nonzero diagonal elements.

*Proof:* See Appendix C. The proof is similar to the proof given in Packard & Doyle (1993) related to the state space upper bound test in $\mu$ theory.

□

**Theorem 5.7 [$l_2$ theory - Diagonal dominance].** Given the representation (5.26), if there exist matrices $P_i$, $N_i$ such that

$$c_\alpha \beta_P < 1 \tag{5.30}$$

with $c_\alpha = \prod_{i=1}^{q}(1+\alpha_i)^{1/2}$ and $\beta_P = ||P_{tot} R_{tot} P_{tot}^{-1}||_2^q$, then

1. For a finite time horizon $N$:

$$\sum_{k=0}^{N-1} ||e_k||_2^2 \le c_\alpha^2 \beta_P^2 \sum_{k=0}^{N-1} ||w_k||_2^2 + r_0^2 \tag{5.31}$$

   where $r_0 = ||D_1 p_0||_2$.

2. For $N \to \infty$:
   There exist constants $c_1, c_2$ such that

$$c_2(1 - c_\alpha^2 \beta_P^2)||p||_2^2 + ||e||_2^2 \le c_\alpha^2 \beta_P^2 ||w||_2^2 + c_1 ||p_0||_2^2 \tag{5.32}$$

   provided that $\{w_k\}_{k=0}^{\infty} \in l_2$. Hence also

$$||e||_2^2 \le c_\alpha^2 \beta_P^2 ||w||_2^2 + c_1 ||p_0||_2^2$$

   holds in that case.

Here the matrix $R_{tot}$ is given by (5.27) and $P_{tot}$ by:

- $P_{tot} = \text{blockdiag}\{P_2, P_3, ..., P_q, P_{S_1}\}$, $P_{S_1} = \text{blockdiag}\{P_1, I_{n_w}\}$
- $Q_i = P_i^T P_i N_i$ are diagonally dominant matrices with $\delta_{Q_i} = (1+\alpha_i)/\alpha_i \ge 1$, $(\alpha_i \ge 0)$, $N_i$ are diagonal matrices with positive diagonal elements and $P_i$ are of full rank.

*Proof:* See Appendix C.

□

**Remark.** The case $N \to \infty$ only applies to systems without bias vectors, because the bias vectors correspond to a constant input 1 in the exogenous input $w_k$. This constant input does not belong to the class $l_2$. The finite $N$ case however is still applicable.

**Theorem 5.8 [$l_2$ theory - Condition number factor].** Given the representation (5.26), if there exist matrices $P_i$ such that

$$c_P \beta_P < 1 \tag{5.33}$$

with $c_P = \prod_{i=1}^{q} \kappa(P_i)$ and $\beta_P = ||P_{tot} R_{tot} P_{tot}^{-1}||_2^q$, then

1. For a finite time horizon $N$:

$$\sum_{k=0}^{N-1} \|e_k\|_2^2 \leq c_P^2 \beta_P^2 \sum_{k=0}^{N-1} \|w_k\|_2^2 + r_0^2 \tag{5.34}$$

where $r_0 = \|D_1 p_0\|_2$.

2. For $N \to \infty$:
There exist constants $c_1, c_2$ such that

$$c_2(1 - c_P^2 \beta_P^2)\|p\|_2^2 + \|e\|_2^2 \leq c_P^2 \beta_P^2 \|w\|_2^2 + c_1 \|p_0\|_2^2 \tag{5.35}$$

provided that $\{w_k\}_{k=0}^{\infty} \in l_2$. Hence also

$$\|e\|_2^2 \leq c_P^2 \beta_P^2 \|w\|_2^2 + c_1 \|p_0\|_2^2$$

holds in that case.

The matrix $R_{tot}$ is given by (5.27) and $P_{tot} = \text{blockdiag}\{P_2, P_3, ..., P_q, P_{S_1}\}$ with $P_{S_1} = \text{blockdiag}\{P_1, I_{n_w}\}$ and $P_i$ full rank matrices.

*Proof:* See Appendix C.

□

**Remark.**

For a linear control system $\Xi_0^0$ Theorem 5.7 or 5.8 is related to the $H_\infty$ control problem with dynamic output feedback (see e.g. Stoorvogel (1992)). In this case is $\Omega_1 = I$ and then it follows immediately from the proof of the Theorem that condition (5.30) and (5.33) become

$$\| \begin{bmatrix} P_1 & 0 \\ 0 & I \end{bmatrix} R_1 \begin{bmatrix} P_1 & 0 \\ 0 & I \end{bmatrix}^{-1} \|_2 < 1. \tag{5.36}$$

Now this is precisely a condition like the one that can be found in the literature in the context of the $H_\infty$ norm for a linear dynamical system (see e.g. Packard & Doyle (1993)). There it is stated that for a linear system

$$\begin{cases} p_{k+1} &= A_R\, p_k + B_R\, w_k \\ e_k &= C_R\, p_k + D_R\, w_k \end{cases}$$

the following five conditions are equivalent to condition (5.36), assuming $R_1 = [A_R\ B_R; C_R\ D_R]$ and the system is stable ($\rho(A_R) < 1$):

1. $\|D_R + C_R(zI - A_R)^{-1}B_R\|_\infty < 1$
2. $S$ has no eigenvalues on the unit circle and $\|D_R + C_R(I - A_R)^{-1}B_R\|_2 < 1$
3. $\exists X \geq 0$ solution to the Riccati equation

$$E^T XE - X - E^T XG(I + XG)^{-1}XE + Q = 0$$

with $I - D_R^T D_R - B_R^T X B_R > 0$ and $(I + GX)^{-1}E$ stable.
4. $\exists X > 0$ such that

$$E^T XE - X - E^T XG(I + XG)^{-1}XE + Q < 0$$

and $I - D_R^T D_R - B_R^T X B_R > 0$.
5. $\exists X > 0$ such that

$$R_1^T \begin{bmatrix} X & 0 \\ 0 & I \end{bmatrix} R_1 - \begin{bmatrix} X & 0 \\ 0 & I \end{bmatrix} < 0$$

where by definition

$$\begin{aligned} E &= A_R + B_R(I - D_R^T D_R)^{-1} D_R^T C_R \\ G &= -B_R(I - D_R^T D_R)^{-1} B_R^T \\ Q &= C_R^T (I - D_R^T D_R)^{-1} C_R. \end{aligned}$$

It is supposed that $E$ is nonsingular and a symplectic matrix $S$ is defined as

$$S = \begin{bmatrix} E + GE^{-T}Q & -GE^{-T} \\ -E^{-T} & E^{-T} \end{bmatrix}.$$

A comparison between Theorem 5.1 and 5.6, 5.2 and 5.7, 5.3 and 5.8, shows that there is a close connection between internal stability (autonomous case) and the property of finite $L_2$-gain. The latter becomes clear through the concept of dissipativity. This was already stated e.g. in Hill & Moylan (1980), Willems (1971)(1972) and van der Schaft (1992) (but in a continuous time framework). The following definition related to discrete time systems is taken from Byrnes & Lin (1994).

**Definition 5.6 [Dissipativity].** A dynamic system with input $w_k$ and output $e_k$ and state vector $p_k$ is called *dissipative* if there exists a nonnegative function $V(p) : \mathbb{R}^{n_p} \to \mathbb{R}$ with $V(0) = 0$, called the *storage function*, such that $\forall w \in \mathbb{R}^{n_w}$ and $\forall k \geq 0$:

$$V(p_{k+1}) - V(p_k) \leq W(e_k, w_k) \tag{5.37}$$

where $W(e_k, w_k)$ is called the *supply rate.*

**Lemma 5.5.** The $NL_q$ system (5.5) is dissipative under the condition of Theorem 5.6, with storage function $V(p) = \|D_1 p\|_2^2$, supply rate $W(e_k, w_k) = \beta_D^2 \|w_k\|_2^2 - \|e_k\|_2^2$ with finite $L_2$-gain $\beta_D < 1$ or under the condition of Theorem 5.7 with storage function $V(p) = \|P_1 p\|_2^2$, supply rate $W(e_k, w_k) = c_\alpha^2 \beta_P^2 \|w_k\|_2^2 - \|e_k\|_2^2$ with finite $L_2$-gain $c_\alpha \beta_P < 1$ or finally under the condition of Theorem 5.8 with storage function $V(p) = \|P_1 p\|_2^2$, supply rate $W(e_k, w_k) = c_P^2 \beta_P^2 \|w_k\|_2^2 - \|e_k\|_2^2$ with finite $L_2$-gain $c_P \beta_P < 1$.

*Proof:* See Appendix C.

□

**Remark.** Several kinds of dissipativity exist in general, depending on the supply rate. A system with supply rate $W = e_k^T w_k$ is called passive, while a supply rate $W = \gamma^2 \|w_k\|_2^2 - \|e_k\|_2^2$ stands for finite $L_2$-gain $\gamma$ (see Hill & Moylan (1976)(1977)(1980)).

## 5.5 Robust performance problem

In the previous Section we have considered sufficient conditions for I/O stability of $NL_q$s. Now we will investigate the influence of parametric uncertainties upon a nominal *nonlinear* $NL_q$ system. This is in fact an extension to the common way of thinking in modern robust control theory, where nominal *linear* systems are defined and uncertainties upon this linear model are analyzed. Conditions for robust stability and robust performance of the $NL_q$ with respect to parametric uncertainties are derived. The main goal of this Section is in fact to show then in what sense and under what conditions $\mu$ control theory (see e.g. Packard & Doyle (1993)) can be considered as a special case of $NL_q$ theory.

### 5.5.1 Perturbed $NL_q$s

Let us assume a perturbation of the form $R_l = R_l^{nom} + \Delta R_l$ upon the $NL_q$ system (5.26), in order to study the influence of changes around the nominal values of the weights on the stability and the $L_2$-gain of the system

$$\begin{bmatrix} p_{k+1} \\ e_k^{ext} \end{bmatrix} = \left(\prod_{i=1}^{l-1} \Omega_i R_i \, \Omega_l (R_l^{nom} + \Delta R_l) \prod_{j=l+1}^{q} \Omega_j R_j\right) \begin{bmatrix} p_k \\ w_k \end{bmatrix}. \tag{5.38}$$

Suppose this perturbation on $R_l$ is through one of the interconnection matrices $X \in \mathbb{R}^{n\times m}$ and that the elements $x_{ij}$ belong to bounded intervals: $x_{ij} \in [\underline{x}_{ij}, \overline{x}_{ij}]$. According to Steinbuch *et al.* (1992), by defining

$$\begin{aligned} x_{ij}^{nom} &= (\overline{x}_{ij} + \underline{x}_{ij})/2 \\ r_{ij} &= (\overline{x}_{ij} - \underline{x}_{ij})/2 \end{aligned} \tag{5.39}$$

one obtains

$$x_{ij} = x_{ij}^{nom} + \delta_{ij} r_{ij},\ \delta_{ij} \in [-1, 1].$$

The matrix $X = X^{nom} + \Delta X$ can be written with

$$\Delta X = S_x \Delta_x R_x \tag{5.40}$$

where $\Delta_x \in \mathbb{R}^{nm\times nm}$ is a real diagonal matrix satisfying the property $||\Delta_x|| \leq 1$ and

$$\begin{aligned} \Delta_x &= \mathrm{diag}\{\delta_{11}, ..., \delta_{1m}, \delta_{21}, ..., \delta_{2m}, \delta_{n1}, ..., \delta_{nm}\} \\ S_x &= \mathrm{blockdiag}\{1_{1\times m}, ..., 1_{1\times m}\} \\ R_x &= [\mathrm{diag}\{r_{1i}\}; \mathrm{diag}\{r_{2i}\}; ...; \mathrm{diag}\{r_{ni}\}],\ (i = 1, ..., m). \end{aligned} \tag{5.41}$$

In order to fix the ideas for a $2 \times 2$ matrix $X$ (5.40) and (5.41) becomes

$$\begin{bmatrix} x_{11} & x_{12} \\ x_{21} & x_{22} \end{bmatrix} = \begin{bmatrix} x_{11}^{nom} & x_{12}^{nom} \\ x_{21}^{nom} & x_{22}^{nom} \end{bmatrix} + \begin{bmatrix} 1 & 1 & 0 & 0 \\ 0 & 0 & 1 & 1 \end{bmatrix} \begin{bmatrix} \delta_{11} & & & \\ & \delta_{12} & & \\ & & \delta_{21} & \\ & & & \delta_{22} \end{bmatrix} \begin{bmatrix} r_{11} & 0 \\ 0 & r_{12} \\ r_{21} & 0 \\ 0 & r_{22} \end{bmatrix}$$

Suppose now that there exist matrices $U_{R_l}$, $V_{R_l}$ such that $R_l(X)$ with a perturbed $X$ can be written as

$$R_l(X) = R_l^{nom} + U_{R_l} \Delta_x V_{R_l}. \tag{5.42}$$

The perturbed system (5.38) can then be written as

$$\left\{ \begin{aligned} \begin{bmatrix} p_{k+1} \\ e_k^{ext} \\ s_k \end{bmatrix} &= \left(\prod_{i=1}^q \Upsilon_i T_i\right) \begin{bmatrix} p_k \\ w_k \\ r_k \end{bmatrix} \\ r_k &= \Delta_x\, s_k, \qquad ||\Delta_x|| \leq 1 \end{aligned} \right. \tag{5.43}$$

with

$$\Upsilon_i = \begin{bmatrix} \Omega_i & \\ & I \end{bmatrix}, T_i = \begin{bmatrix} R_i & \\ & I \end{bmatrix} (i \neq l), T_l = \begin{bmatrix} R_l^{nom} & U_{R_l} \\ V_{R_l} & 0 \end{bmatrix}$$

where $\|\Upsilon_i\| \leq 1 \quad (i = 1, ..., q)$. As stated in the following Lemma, such a perturbed $\mathrm{NL}_q$ is nothing else but an $\mathrm{NL}_{q+1}$.

**Lemma 5.6 [Perturbed $\mathbf{NL}_q$s as $\mathbf{NL}_{q+1}$s].** The perturbed $\mathrm{NL}_q$ (5.43) can be written as an $\mathrm{NL}_{q+1}$ system with $T_{q+1} = I$ and $\Upsilon_{q+1} = \mathrm{diag}\{I, I, \Delta_x\}$ as

$$\begin{bmatrix} p_{k+1} \\ e_k^{ext} \\ s_k \end{bmatrix} = (\prod_{i=1}^{q} \Upsilon_i T_i \begin{bmatrix} I & & \\ & I & \\ & & \Delta_x \end{bmatrix} I) \begin{bmatrix} p_k \\ w_k \\ s_k \end{bmatrix} \tag{5.44}$$

with $\Delta_x$ diagonal and $\|\Delta_x\|_2 \leq 1$.

*Proof:* Eliminate $r_k$ in (5.43).

□

The following example illustrates the perturbed $\mathrm{NL}_q$ (5.43).

**Example 5.3.** The $\Xi_1^0$ system (autonomous case) with uncertainties on the elements of $C$ ($C = C^{nom} + \Delta C$ with $\Delta C = S_c \Delta_c R_c$) is:

$$\begin{bmatrix} x_{k+1} \\ z_{k+1} \\ \xi_{k+1} \end{bmatrix} = \Gamma_1 \begin{bmatrix} A & 0 & B \\ F(C+\Delta C) & E & FD \\ GF(C+\Delta C) & GE & GFD \end{bmatrix} \begin{bmatrix} x_k \\ z_k \\ \xi_k \end{bmatrix}$$

or in the form (5.43)

$$\begin{cases} \left[\begin{array}{c} x_{k+1} \\ z_{k+1} \\ \xi_{k+1} \\ \hline s_k \end{array}\right] = \left[\begin{array}{c|c} \Gamma_1 & 0 \\ \hline 0 & I \end{array}\right] \left[\begin{array}{ccc|c} A & 0 & B & 0 \\ FC & E & FD & FS_c \\ GFC & GE & GFD & GFS_c \\ \hline R_c & 0 & 0 & 0 \end{array}\right] \left[\begin{array}{c} x_k \\ z_k \\ \xi_k \\ \hline r_k \end{array}\right] \\ r_k = \Delta_c s_k, \qquad \|\Delta_c\| \leq 1 \end{cases}$$

with

$$U_{R_l} = \begin{bmatrix} 0 \\ FS_c \\ GFS_c \end{bmatrix}, V_{R_l} = [R_c \;\; 0 \;\; 0] \; (l = 1)$$

which is easily verified by straightforward calculation.

The following Theorem holds then for input/output stability of the perturbed $\mathrm{NL}_q$.

**Theorem 5.9 [$l_2$ theory - perturbed NL$_q$ system].**
If there exist matrices $P_i, N_i$ such that

$$c_\alpha \beta_P < 1 \tag{5.45}$$

with $c_\alpha = \prod_{i=1}^{q}(1+\alpha_i)^{1/2}$ and $\beta_P = \|P_{tot}T_{tot}P_{tot}^{-1}\|_2^q$, then the perturbed NL$_q$ system (5.43) is I/O stable for all real diagonal matrices $\Delta_x$ that satisfy the condition $\|\Delta_x\| \leq 1$:

$$\|e\|_2^2 \leq c_\alpha^2 \beta_P^2 \|w\|_2^2 + c_1\|p_0\|_2^2 \tag{5.46}$$

provided that $\{w_k\}_{k=0}^{\infty} \in l_2$. The matrices $T_{tot}$ and $P_{tot}$ are given by:

–

$$T_{tot} = \begin{bmatrix} 0 & T_2 & & & 0 \\ & 0 & T_3 & & \\ & & \ddots & & \\ & & & 0 & T_q \\ T_1 & & & & 0 \end{bmatrix}, T_i \in \mathbb{R}^{n_{t_i} \times n_{t_{i+1}}}, n_{t_1} = n_{t_{q+1}}$$

- $P_{tot} = \text{blockdiag}\{P_2, P_3, ..., P_q, P_{S_1}\}$, $P_{S_1} = \text{blockdiag}\{P_1, I, D_1\}$
- $Q_i = P_i^T P_i N_i$ are diagonally dominant matrices with $\delta_{Q_i} = (1+\alpha_i)/\alpha_i \geq 1$ $(\alpha_i \geq 0)$, $N_i$ are diagonal matrices with positive diagonal elements and $P_i$ of full rank.
- $D_1$ is diagonal with nonzero diagonal elements.

Furthermore if there exist a matrix $P_{S_1}$ and matrices $P_i$ $(i = 2, ..., q)$ such that

$$c_P \beta_P < 1 \tag{5.47}$$

with $c_P = \kappa(P_{S_1}) \prod_{i=2}^{q} \kappa(P_i)$, then

$$\|e\|_2^2 \leq c_P^2 \beta_P^2 \|w\|_2^2 + c_1\|p_0\|_2^2 \tag{5.48}$$

holds for all real diagonal matrices $\Delta_x$ that satisfy $\|\Delta_x\| \leq 1$.

*Proof:* See Appendix C.

□

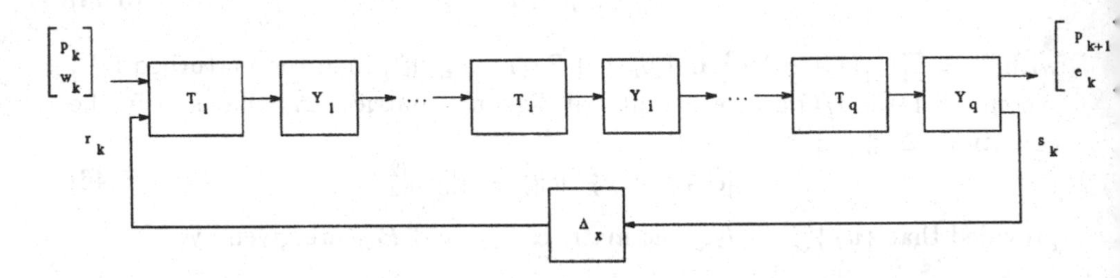


Figure 5.7: *Perturbed $NL_q$ system where parametric uncertainties upon a given matrix of the $NL_q$ are taken into account through the feedback perturbation $\Delta_x$.*

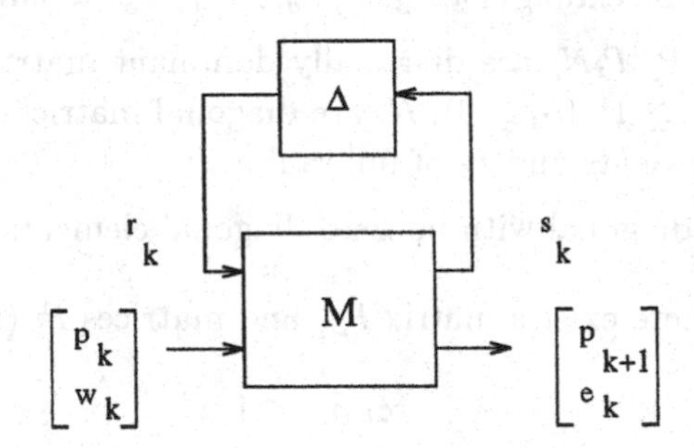


Figure 5.8: *LFT (Linear Fractional Transformation) representation as it occurs in modern robust control theory ($\mu$ theory). $M$ is the augmented plant (nominal linear system) and the uncertainty on $M$ is represented through the feedback interconnection $\Delta$. Remark the similarity between this LFT and the perturbed $NL_q$ of Figure 5.7. For real parametric uncertainties a perturbed $NL_1$ reduces to an LFT.*

### 5.5.2 Connections with $\mu$ theory

In $\mu$ theory (see e.g. Packard & Doyle (1993), Maciejowski (1989)) one analyses the influence of parametric uncertainties, unmodeled dynamics etc. related to a nominal linear model on the system's performance or one takes into account these uncertainties in the controller synthesis problem. Uncertainty is formulated then by means of LFTs (Linear Fractional Transformations). Related to the standard plant a state space formulation for an LFT looks like (see Figure 5.8)

$$\left\{ \begin{array}{rcl} \begin{bmatrix} x_{k+1} \\ e_k^{ext} \\ s_k \end{bmatrix} & = & M \begin{bmatrix} x_k \\ w_k \\ r_k \end{bmatrix} \\ r_k & = & \Delta \, s_k \end{array} \right. \tag{5.49}$$

with $M = [M_{11}\ M_{12}\ M_{13}; M_{21}\ M_{22}\ M_{23}; M_{31}\ M_{32}\ M_{33}]$. In general a block uncertainty structure is considered for $\Delta \in \mathbf{\Delta} \subset \mathbb{C}^{n_{tot} \times n_{tot}}$ (see Doyle (1982), Packard & Doyle (1993))

$$\begin{array}{rl} \mathbf{\Delta} = \{\text{diag}[\delta_1 I_{r_1}, ..., \delta_s I_{r_s}, \Delta_{s+1}, ..., \Delta_{s+f}] : & \delta_i \in \mathbb{C}, \Delta_{s+j} \in \mathbb{C}^{m_j \times m_j}; \\ & i = 1, ..., s; j = 1, ..., f\} \end{array} \tag{5.50}$$

where $\sum_{i=1}^{s} r_i + \sum_{j=1}^{f} m_j = n_{tot}$ for consistency. Remark that these uncertainty blocks are not necessarily real, but can take complex values. The fact that these are real or complex depends on the specific problem that one studies: parametric uncertainties are real in nature, while unmodeled dynamics may lead to complex values. The uncertainties can be of time-invariant, time-variant or nonlinear nature. For linear time-invariant uncertainties $\Delta$, a necessary and sufficient condition for robust performance (finite $L_2$-gain smaller than one, regardless of any $\Delta$ belonging to $\mathbf{B}_{\mathbf{\Delta}} = \{\Delta \in \mathbf{\Delta} : \|\Delta\|_2 \leq 1\}$):

$$\|e\|_2 \leq \beta \|w\|_2 \qquad \beta \in [0, 1) \tag{5.51}$$

for zero-initial-state-response is

$$\mu_{\Delta_S}(M) < 1 \tag{5.52}$$

where $\Delta_S$ is an augmented block structure for $\Delta$ and $\mu_{\Delta_S}(M)$ denotes the *structured* singular value of the matrix $M$ related to the uncertainty block structure $\Delta_S$. Some properties of $\mu$ are

$$\rho(M) \leq \max_{Q \in \mathcal{Q}} \rho(QM) \leq \max_{\Delta \in \mathbf{B}_{\mathbf{\Delta}}} \rho(\Delta M) = \mu_\Delta(M) \leq \min_{D \in \mathcal{D}} \|DMD^{-1}\|_2 \leq \|M\|_2 \tag{5.53}$$

where $\mathcal{Q} = \{Q \in \Delta : Q^*Q = I_n\}$, $\mathcal{D} = \{\text{diag}[D_1, ..., D_s, d_{s+1}I_{m_1}, ..., d_{s+f}I_{m_f}] : D_i \in \mathbb{C}^{r_i \times r_i}, D_i = D_i^* > 0, d_{s+j} \in \mathbb{R}, d_{s+j} > 0\}$.

The upper bound of $\mu$ in (5.53) corresponds to diagonal scaling of $M$, which is similar to the diagonal scaling in Theorem 5.1 or 5.6. Working with this upper bound instead of $\mu$, guarantees also robust performance with respect to uncertainties of nonlinear and time-varying nature (the sufficient but not necessary condition in that case might be too conservative). The following Theorem is due to Packard and Doyle.

**Theorem 5.10 [Packard & Doyle].** Given the LFT (5.49), if

$$\min_{D_S} \|D_S M D_S^{-1}\|_2 = \beta < 1 \tag{5.54}$$

with $D_S = \text{diag}\{D_1, d_2 I, D\}$ and $D_1, D$ diagonal matrices with positive elements, then the following holds for the zero initial state

$$\|e\|_2^2 \leq \beta^2 \|w\|_2^2 \tag{5.55}$$

if $\{w_k\}_{k=0}^{\infty} \in l_2$.

*Proof:* See Packard & Doyle (1993).

□

We have already mentioned the similarity between LFTs and perturbed $NL_q$s. The following Theorem makes a more precise statement. It states that for a real diagonal $\Delta$ uncertainty block the Theorem by Packard & Doyle is a special case of Theorem 5.9.

**Theorem 5.11 [Connection between Theorem 5.9 and 5.10].** The LFT (5.49) is a special case of the perturbed $NL_q$ system (5.43) and Theorem 5.10 a special case of Theorem 5.9 under the following conditions

1. $q = 1$ and $\Upsilon_1 = I$
2. $\delta_{Q_i} \to \infty$ (Diagonal scaling case)
3. $M_{33} = 0$
4. The uncertainty block $\Delta$ is real
5. The diagonal $\Delta$ block satisfies $r_i = 0 (\forall i)$, $m_j = 1 (\forall j)$, $s = 0$, $f = n_{tot}$ in (5.50).

*Proof:* Readily checked by comparing (5.43)(5.45) with (5.49)(5.50)(5.54).

□

Hence the state space upper bound test in $\mu$ theory with constant $D$ scaling is closely related to the diagonal scaling criteria in $\text{NL}_q$ theory.

## 5.6 Stability analysis: formulation as LMI problems

Now we show how the $\text{NL}_q$ stability criteria can be cast as Linear Matrix Inequality (LMI) problems. These problems are convex which means they have a unique minimum and moreover this minimum can be found in polynomial time.

Let us first summarize the $\text{NL}_q$ stability criteria here:

- *Diagonal scaling:* find diagonal matrices $D_i$ such that

$$\|D_{tot} Z_{tot} D_{tot}^{-1}\|_2^q < 1. \tag{5.56}$$

- *Diagonal dominance:* find matrices $P_i$ and diagonal matrices $N_i$ such that

$$c_\alpha \|P_{tot} Z_{tot} P_{tot}^{-1}\|_2^q < 1 \tag{5.57}$$

and

$$c_\alpha = \prod_{i=1}^{q} (1+\alpha_i)^{1/2}, \quad \alpha_i = \frac{1}{\delta_{Q_i} - 1}, \quad Q_i = P_i^T P_i N_i.$$

- *Condition number factor:* find matrices $P_i$ such that

$$c_P \|P_{tot} Z_{tot} P_{tot}^{-1}\|_2^q < 1 \tag{5.58}$$

with

$$c_P = \kappa(P_{S_1}) \prod_{i=2}^{q} \kappa(P_i).$$

- *Choice of $Z_{tot}, D_{tot}, P_{tot}$:*

$$\begin{aligned} Z_{tot} &= V_{tot} &&: \text{ global asymptotic stability} \\ &= R_{tot} &&: \text{ finite } L_2\text{-gain} < 1 \\ &= T_{tot} &&: \text{ robust performance} \end{aligned}$$

and $D_{tot} = \text{diag}\{D_2, D_3, ..., D_q, D_{S_1}\}$ with

$$\begin{aligned} D_{S_1} &= D_1 && : \text{global asymptotic stability} \\ &= \text{diag}\{D_1, I\} && : \text{finite } L_2\text{-gain} < 1 \\ &= \text{diag}\{D_{1_a}, I, D_{1_b}\} && : \text{robust performance} \end{aligned}$$

and $P_{tot} = \text{blockdiag}\{P_2, P_3, ..., P_q, P_{S_1}\}$ with

$$\begin{aligned} P_{S_1} &= P_1 && : \text{global asymptotic stability} \\ &= \text{blockdiag}\{P_1, I\} && : \text{finite } L_2\text{-gain} < 1 \\ &= \text{blockdiag}\{P_1, I, D_1\} && : \text{robust performance} \end{aligned}$$

The problems (5.56)-(5.58) correspond to nondifferentiable optimization problems that involve matrix inequalities of the form $M(x) < 0$ with $M = M^T$ with unknown $x \in \mathbb{R}^n$ because an expression like $\|A\|_2 < \gamma$ is equivalent to $A^T A - \gamma I < 0$. For the special case of linear matrix inequalities (LMIs) this matrix $M$ depends affinely on $x$: $M(x) = M_0 + \sum_{i=1}^{m} x_i M_i$ with $M_0 = M_0^T$, $M_i = M_i^T$. A set of LMIs $M_l(x) < 0$ $(l = 1, ..., L)$ is equivalent to a single LMI by considering $M = \text{blockdiag}\{M_1, ..., M_L\} < 0$ (see Boyd *et al.* (1994)). Finding a feasible $x$ to such an LMI is a convex problem, which has a unique minimum. LMIs occur frequently in systems and control problems. The recent book by Boyd *et al.* (1994) gives an overview on this subject. From a computational point of view LMIs are attractive because efficient algorithms exist for solving them. A general theory of interior-point polynomial-time methods for convex programming is presented in Nesterov & Nemirovskii (1994): this area has been intensively developed since Karmarkar's famous paper in 1984 on linear programming (Karmarkar (1984)). Other related work is Boyd & El Ghaoui (1993), Nemirovskii & Gahinet (1994), Vandenberghe & Boyd (1993), Overton (1988). Software for solving LMIs is available in Matlab's LMI lab (see Gahinet & Nemirovskii (1993)). General type of LMIs can be solved in LMI lab of the form

$$\mathcal{N}^T \mathcal{L}(X_1, ..., X_K) \mathcal{N} \leq \mathcal{M}^T \mathcal{R}(X_1, ..., X_K) \mathcal{M} \tag{5.59}$$

where $X_i$ are the unknown matrix variables, $\mathcal{L}(X_1, ..., X_K)$ and $\mathcal{R}(X_1, ..., X_K)$ are symmetric block matrices, each block defining an affine combination of the matrices $X_i$ and their transpose and $\mathcal{N}$, $\mathcal{M}$ are constant matrices.
Let us discuss the problems (5.56)-(5.58) now in more detail

- **Diagonal scaling:** The LMI which corresponds to (5.56) is

$$Z_{tot}^T D_{tot}^2 Z_{tot} < D_{tot}^2 \tag{5.60}$$

and can be handled in LMI lab. The function *psv* of Matlab's Robust Control Toolbox (The MathWorks Inc. (1994)) computes $\min_{D_{tot}} \|D_{tot} Z_{tot} D_{tot}^{-1}\|_2$ for the case $D_{S_1} = D_1$ (global asymptotic stability).

- **Diagonal dominance:** A set of LMIs will be associated now to the diagonal dominance case (5.57). In the sequel we have to assume that $N_i = I$ for $i = 1, ..., q$. The condition is formulated then as a convex feasibility problem in the unknown matrices $Q_i$, $Q_{S_1}$

$$Z_{tot}^T Q_{tot} Z_{tot} < \gamma_1 Q_{tot} \tag{5.61}$$

with $Q_{tot} = P_{tot}^T P_{tot} = \text{blockdiag}\{Q_2, ..., Q_q, Q_{S_1}\}$, $Q_i = P_i^T P_i$, $Q_{S_1} = P_{S_1}^T P_{S_1}$, such that

$$\delta_{Q_i} > \gamma_{2_i} \ (\forall i \mid \Gamma_i \neq I), \tag{5.62}$$

where $\gamma_1 \leq 1$, $\gamma_{2_i} \geq 1$ are user defined constants. In order to obtain a set of LMIs the constraint for diagonal dominance is replaced by

$$\left\{ \begin{array}{rcl} Q_i - \gamma_{3_i} I & > & 0 \ (\forall i \mid \Gamma_i \neq I) \\ \|Q_i\|_2 & < & \gamma_{4_i} \ (\forall i \mid \Gamma_i \neq I) \\ Q_i & > & 0, \ (\forall i) \end{array} \right. \tag{5.63}$$

and the Schur complement form is used for $\|Q_i\|_2 < \gamma_{4_i}$ which is

$$\begin{bmatrix} \gamma_{4_i} I & Q_i \\ Q_i^T & \gamma_{4_i} I \end{bmatrix} > 0. \tag{5.64}$$

A qualitative explanation why the set of LMIs (5.63) enforces the matrices $Q_i$ to be diagonally dominant can be given based on Gerschgorin's Theorem and Lemma 5.3. Let us fix the ideas for $q = 1$. Gerschgorin's Theorem applied to the matrix $Q - \gamma_3 I$ states that the discs, containing the eigenvalues of the matrix, are centered at $q_{ii} - \gamma_3$ with radii $\sum_{j(j \neq i)} |q_{ij}|$. Now it is clear that for $\gamma_3 > 0$ and if $Q - \gamma_3 I > 0$ holds an increasing value $\gamma_3$ enforces $Q$ to be more diagonally dominant. The introduction of an upper bound on $\|Q\|_2$ follows from Lemma 5.3. Indeed because $Q = D_Q(I + X_Q)$ and $\|X_Q\|_2 \leq \sqrt{n}\|X_Q\|_\infty$ (assuming $Q \in \mathbb{R}^{n \times n}$) and $\|X_Q\|_\infty < 1/\delta_Q$ one obtains

$$\|Q\|_2 < \|D_Q\|_2(1 + \sqrt{n}/\delta_Q) \tag{5.65}$$

which means that some upper bound $\gamma_4$ on $\|Q\|_2$ is needed. The choice of $\gamma_3$ in (5.63) influences $\|D_Q\|_2$. One may use the following rule of thumb: choose first $\gamma_3$ and secondly $\gamma_4$ proportionally to $\gamma_3$ (e.g. $\gamma_4 = 10\gamma_3$). This was affirmed by computer simulations using LMI lab, by trying several combinations of $\gamma_1, \gamma_3, \gamma_4$ on random matrices $Z_{tot}$.

- **Condition number factor:** In this case it is much easier to formulate a set of LMIs than in the diagonal dominance case. Indeed according to Boyd *et al.* (1994) an upper bound on the condition number of a matrix $P$: $\kappa(P) < \alpha$ can be formulated as the LMI

$$I < P^T P < \alpha^2 I.$$

Hence the problem can be interpreted as a feasibility problem in the matrices $Q_i$, $Q_{S_1}$

$$\begin{cases} Z_{tot}^T Q_{tot} Z_{tot} < \gamma^2 Q_{tot} \\ I < Q_{S_1} < \beta^2 I \\ I < Q_i < \alpha_i^2 I, \quad i = 2, ..., q \end{cases} \tag{5.66}$$

where $Q_{tot} = P_{tot}^T P_{tot} = \text{blockdiag}\{Q_2, ..., Q_q, Q_{S_1}\}$ and $Q_i = P_i^T P_i$, $Q_{S_1} = P_{S_1}^T P_{S_1}$. In order to find a feasible point to the set of LMIs, one chooses $\gamma < 1$ and tries to make $\beta$ and $\alpha_i$ as small as possible.

## 5.7 Neural control design

In the previous Sections sufficient conditions for global asymptotic stability and finite $L_2$-gain of $NL_q$s have been derived. Moreover it has been shown that these criteria can be formulated as LMI problems, which are computationally feasible. However the final goal is to synthesize a neural controller, given some identified neural state space model.

The idea is now basically to combine the existing paradigms of classical neural control theory and modern robust control theory:

1. The dynamic backpropagation paradigm, which can be applied in order to let the controller learn to track a specific reference input.

2. The desirable properties of (robust) internal stability and (robust) performance as they are formulated in modern control theory (see e.g. Boyd & Barratt (1991), Maciejowski (1989)).

The tracking of a specific reference input using dynamic backpropagation has already been explained in Chapter 4. The following cost function has been minimized for a specific reference input $d_k$:

$$\min_{\theta_c} J(\theta_c) = \frac{1}{2N} \sum_{k=1}^{N} \{[d_k - \hat{y}_k(\theta_c)]^T [d_k - \hat{y}_k(\theta_c)] + \lambda . u_k(\theta_c)^T u_k(\theta_c)\}. \tag{5.67}$$

The dynamic backpropagation procedure will then be modified with a constraint that imposes internal or I/O stability or robust performance. In that case one can consider inputs to the system that belong to the class $l_2$, instead of specific inputs only.

## 5.7.1 Synthesis problem

More difficult than the analysis problem is the synthesis problem where the controller parameter vector $\theta_c$ belongs to the unknowns of the optimization problem. In general this leads to non-convex nondifferentiable optimization problems. We have the following two types of synthesis problems

1. **Type 1:**

   - *Diagonal scaling:* feasibility problem in $\theta_c$, $D_i$ such that

   $$Z_{tot}(\theta_c)^T D_{tot}^2 Z_{tot}(\theta_c) < D_{tot}^2. \tag{5.68}$$

   - *Diagonal dominance:* feasibility problem in $\theta_c$, $Q_i = Q_i^T$ such that

   $$\left\{ \begin{array}{c} Z_{tot}(\theta_c)^T Q_{tot} Z_{tot}(\theta_c) < \gamma_1 Q_{tot} \\ Q_i - \gamma_{3_i} I > 0 \\ Q_i > 0 \\ \begin{bmatrix} \gamma_{4_i} I & Q_i \\ Q_i^T & \gamma_{4_i} I \end{bmatrix} > 0 \end{array} \right. \tag{5.69}$$

   where $\gamma_1, \gamma_{3_i}, \gamma_{4_i}$ are user defined constants and $Q_i > 0$ must hold $\forall i$ and the other conditions $\forall i$ for which $\Gamma_i \neq I$.

   - *Condition number factor:* feasibility problem in $\theta_c$, $Q_i = Q_i^T$ such that

   $$\left\{ \begin{array}{rcl} Z_{tot}(\theta_c)^T Q_{tot} Z_{tot}(\theta_c) & < & \gamma^2 Q_{tot} \\ I < Q_{S_1} & < & \beta^2 I \\ I < Q_i & < & \alpha_i^2 I, \quad i = 2, ..., q \end{array} \right. \tag{5.70}$$

   with $\alpha_i$, $\beta$, $\gamma$ user defined constants.

2. **Type 2:** feasible controller $\theta_c$ with optimal tracking of a given reference input

   $$\min_{\theta_c, Q_i \text{ or } D_i} J(\theta_c) \tag{5.71}$$

   such that the type 1 condition (diagonal scaling (5.68), diagonal dominance (5.69) or condition number factor (5.70)) holds. The cost function $J$ corresponds to (5.67).

### 5.7.2 Non-convex nondifferentiable optimization

For a fixed value of $\theta_c$ the type 1 problems (5.68)-(5.70) are convex. However the overall type 1 & 2 problem are non-convex and nondifferentiable. A framework for solving non-convex nondifferentiable optimization problems with singular value inequality constraints was proposed in Polak & Wardi (1982) in the context of semi-infinite optimization and can be used for the synthesis problem. Also ellipsoid algorithms have been applied to non-convex problems, although they are normally intended for convex problems (see e.g. Ecker & Kupferschmid (1985), Shor (1985), Boyd & Barratt (1991)). A disadvantage of all these algorithms is that they are quite slow and in fact a lot of work remains to be done on quadratically convergent algorithms for non-convex nondifferentiable optimization.

An essential aspect in non-differentiable problems is to replace the gradient by a *generalized gradient*. In Polak & Wardi (1982) the generalized gradient is discussed related to the problem

$$\min_{x\in\mathbb{R}^n} \lambda_{max}[M(x)], \qquad M = M^T > 0 \tag{5.72}$$

where $\lambda_{max}$ is the maximal eigenvalue of $M$. $M > 0$ denotes the matrix inequality related to (5.68)-(5.70). Some properties related to $\lambda_{max}(x)$ are that this function is Lipschitz continuous and differentiable whenever $\lambda_1(x) \neq \lambda_2(x)$. In that case the gradient is equal to

$$[\nabla\lambda_{max}[M(x)]]_i = u_1(x)^T \frac{\partial M(x)}{\partial x_i} u_1(x) \tag{5.73}$$

where $u_1(x)$ is the unit eigenvector corresponding to the largest eigenvalue $\lambda_1(x)$ of $M(x)$. In the case of *linear* matrix inequalities the derivative $\frac{\partial M(x)}{\partial x_i}$ corresponds to a constant matrix. At the points where $\lambda_{max}(x)$ is not differentiable a generalized gradient must be considered

$$\partial\lambda_{max}[M(x)] = co\{v \in \mathbb{C}^n \mid v_i = (Uz)^* \frac{\partial M(x)}{\partial x_i}(Uz), z \in \mathbb{C}^{k(x)}, \|z\| = 1\} \tag{5.74}$$

where $co$ denotes the convex hull of the set in $\{.\}$ and $k(x)$ is such that $\lambda_1(x) = \lambda_2(x) = \ldots = \lambda_{k(x)}(x) > \lambda_{k(x)+1}(x)$. One should also note that the use of a generalized gradient doesn't necessarily lead to a descent direction like in differentiable problems. The algorithms described in Polak & Wardi (1982), Polak *et al.* (1984), Polak & Mayne (1985) make use of the (generalized) gradients of the objective function and of the constraints. Such algorithms have been applied to the design of linear multivariable feedback systems e.g. in Polak & Salcudean (1989).

### 5.7.3 A modified dynamic backpropagation algorithm

Instead of applying methods of non-convex nondifferentiable optimization to the general problem in $\theta_c$ and $Q_i$ or $D_i$ it is possible to exploit the fact that the Type 1 or 2 problem contains a ***convex subproblem***. Without loss of generality let us consider the Type 2 problem with diagonal scaling:

$$\min_{\theta_c, D_i} J(\theta_c) \quad \text{such that} \quad M(\theta_c, D_i) < 0. \tag{5.75}$$

This can be thought of as two nested optimization problems: a non-convex and a convex one. For a given value of $\theta_c$, say $\theta_c^0$, the 'inner' feasibility problem

$$M(\theta_c^0, D_i) < 0 \tag{5.76}$$

is convex. The 'outer' optimization problem is then

$$\min_{\theta_c} J(\theta_c) \quad \text{such that} \quad \lambda_{max}[M(\theta_c)] < 0. \tag{5.77}$$

The gradient of $J(\theta_c)$ follows from dynamic backpropagation and the gradient for the constraint at a given point $\theta_c^0$ is given by

$$[\nabla \lambda_{max}[M(\theta_c^0)]]_i = u_1(\theta_c^0)^T \frac{\partial M(\theta_c^0, D_i^{(*)})}{\partial \theta_{c_i}^0} u_1(\theta_c^0) \tag{5.78}$$

with $u_1(\theta_c^0)$ the unit eigenvector corresponding to the largest eigenvalue $\lambda_1(\theta_c^0)$ of $M(\theta_c^0, D_i^{(*)})$ and $D_i^{(*)}$ is the solution to the convex feasibility problem (5.76).

**Remarks.**

- It may happen that it is impossible to find a feasible point $\theta_c^0$ to (5.76), especially in the case of diagonal scaling because the conditions might be conservative. One may not conclude then that the system is not globally asymptotically stable because the condition is only sufficient. In that case one has either to rely on the classical dynamic backpropagation algorithm or one may impose *local* stability of the $NL_q$ at $p = 0$, which leads again to a matrix inequality constraint (see (5.17)). A similar constraint of local stability at a target point was also successfully imposed in order to solve the swinging up problem for an inverted pendulum system in Chapter 4. Moreover, we will show in Section 5.8 that imposing local stability together with making the matrices $Q_i$ as diagonally dominant as possible or making the matrices $P_i$ as well conditioned as possible, leads to good solutions.

- It is also possible to consider a modified dynamic backpropagation algorithm for the system identification problem (Chapter 3) like for the tracking problem. In that case the objective of the prediction error algorithm is modified with an LMI constraint that expresses global asymptotic stability of the model

$$\min_{\theta_m, D_i} V_N(\theta_m, Z^N) \quad \text{such that} \quad M(\theta_m, D_i) < 0. \tag{5.79}$$

## 5.8 Control design: some case studies

### 5.8.1 A tracking example on diagonal scaling

In this example we consider the following nonlinear plant $\mathcal{P}$, neural state space model $\mathcal{M}_2$, linear dynamic controller $\mathcal{C}_0$ and a reference model $\mathcal{L}$:

$$\mathcal{P}: \quad y_{k+1} = 0.3(1+y_k^2)y_k + 0.5u_k$$

$$\mathcal{M}_2: \quad \begin{cases} \hat{x}_{k+1} &= W_{AB}\tanh(V_A\hat{x}_k + V_B u_k + \beta_{AB}) \\ y_k &= W_{CD}\tanh(V_C\hat{x}_k) + \epsilon_k \end{cases}$$

$$\mathcal{C}_0: \quad \begin{cases} z_{k+1} &= Ez_k + Fy_k + F_2 d_k \\ u_k &= Gz_k \end{cases}$$

$$\mathcal{L}: \quad d_{k+1} = 0.6d_k + r_k.$$

First nonlinear system identification was done according to Chapter 3. Input/output data were generated by taking white noise for $u_k$ (uniformly distributed in the interval $[-0.5, 0.5]$). The data set consists of 2000 data points (first 1000 data are the training set and the following 1000 data are the test set). The structure of the neural network model is $n = 1$ and 7 hidden neurons for the state equation and for the output equation and $x_0 = y_0 = 0$. A quasi-Newton method with BFGS updating of the Hessian and mixed quadratic and cubic line search was applied (function *fminu* of Matlab's optimization toolbox (The MathWorks Inc. (1994))). The model and its associated sensitivity model were simulated in C code using Matlab's *cmex* facility. 50 different starting points (randomly chosen according to a normal distribution with zero mean and variance 0.1) for the parameter vector were generated. The selected solution had a fitting error of $V_{fit} = 8.3731\,e-05$ and a generalization error of $V_{gen} = 7.9801\,e-05$ and was acceptable according to correlation tests on the training data and the test data.

Then a tracking problem was specified with $r_k = 0.2\,(\sin(2\pi\, k/25) + \sin(2\pi\, k/10))$ for $k = 1, .., 100$. The closed loop system is an NL$_2$ system with state vector $p_k = [\hat{x}_k; z_k; \xi_k; d_k]$, $\xi_k = \tanh(V_C \hat{x}_k)$, input $w_k = [r_k; \epsilon_k; 1]$

$$\begin{cases} \hat{x}_{k+1} &= W_{AB} \tanh(V_A \hat{x}_k + V_B G z_k + \beta_{AB}) \\ z_{k+1} &= E z_k + F W_{CD} \xi_k + F \epsilon_k + F_2 d_k \\ \xi_{k+1} &= \tanh(V_C W_{AB} \tanh(V_A \hat{x}_k + V_B G z_k + \beta_{AB})) \\ d_{k+1} &= 0.6 d_k + r_k \\ e_k &= d_k - \hat{y}_k \end{cases}$$

and the matrices $V_1, V_2$ of the NL$_q$ are equal to

$$\mathrm{V}_1 = \begin{bmatrix} W_{AB} & & & \\ & I & & \\ & & V_C W_{AB} & \\ & & & 1 \end{bmatrix}, \mathrm{V}_2 = \begin{bmatrix} V_A & V_B G & 0 & 0 \\ 0 & E & F W_{CD} & F_2 \\ V_A & V_B G & 0 & 0 \\ 0 & 0 & 0 & 0.6 \end{bmatrix}.$$

The original dynamic backpropagation algorithm as well as the modified dynamic backpropagation algorithm with global asymptotic stability of the closed loop system (diagonal scaling) were applied. A quasi-Newton method with BFGS updating of the Hessian was used for dynamic backpropagation (function *fminu* of Matlab's optimization toolbox). For the modified dynamic backpropagation algorithm the inner convex optimization problem was solved using the function *psv* of Matlab's robust control toolbox and the outer constrained optimization problem by an SQP method (Sequential Quadratic Programming) using the function *constr* of Matlab's optimization toolbox with numerical calculation of the gradients and simulation of the closed loop system in C code. This SQP method, which is intended for differentiable optimization problems, is meaningful here as long as the largest eigenvalue related to the LMI does not coincide with any other eigenvalue. In the experiments the order of the controller was chosen to be equal to 3 in all cases. The results are shown on Figure 5.9. A solution to the unconstrained problem is plotted on (Top). In (Middle) and (Bottom) a constraint of $\|D_{tot} V_{tot} D_{tot}^{-1}\|_2 < 0.99$ and $< 0.85$ was introduced respectively. Starting from the same random initial controller in the three cases the following minima were obtained in (5.67) and (5.75): J = 0.0066 (Top), J = 0.0073 (Middle), J = 0.0124 (Bottom). Hence the tracking performance for the specific reference input degrades by ensuring a higher contraction rate of the flow for the autonomous control system. The solution corresponding to (Top) is not a feasible point to (5.60).

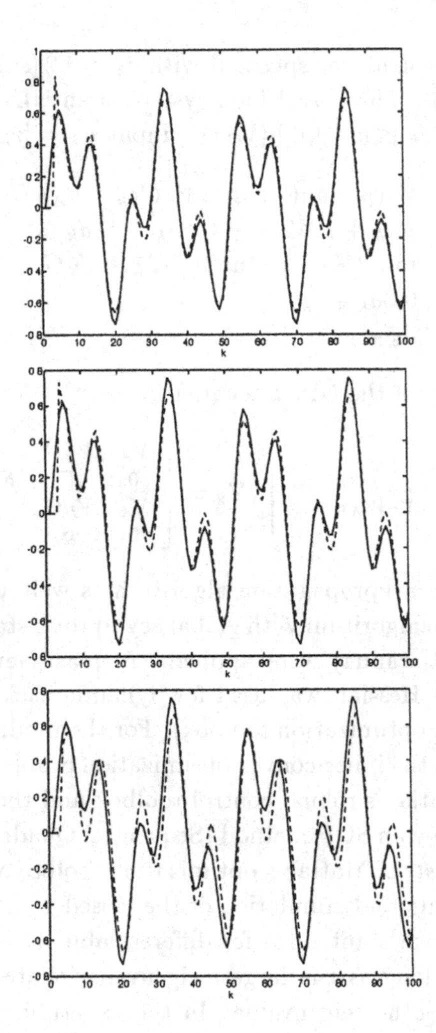


Figure 5.9: *Example on the modified dynamic backpropagation algorithm. Narendra's dynamic backpropagation is modified here with a diagonal scaling condition in order to ensure global asymptotic stability of the model reference control system. (Top) classical dynamic backpropagation: best tracking performance on a specific reference input; (Middle) and (Bottom) modified dynamic backpropagation with guaranteed global asymptotic stability of the closed loop system according to the diagonal scaling case. The tracking performance for the specific reference input degrades for decreasing upper bound on* $\|D_{tot}V_{tot}D_{tot}^{-1}\|_2$*: upper bounds are 0.99 and 0.85 for (Top) and (Bottom) respectively. (Full line: output of reference model, dashed line: output of controlled plant).*

### 5.8.2 A collection of stabilization problems

For neural state models which are themselves already globally asymptotically stable according to the diagonal scaling theorem, it is feasible to find a controller which makes the closed loop globally asymptotically stable according to the same theorem. However for neural state models that reveal other type of behaviour, such as multiple equilibria, periodic behaviour, quasi-periodic behaviour and chaos the diagonal scaling Theorems are useless in order to design stabilizing controllers. The aim of this example is to show that in that case the theorems on diagonal dominance and the theorems with condition number factor can be adopted for designing globally stabilizing controllers.

In order to illustrate this, we consider the autonomous neural state space model

$$x_{k+1} = W_A \tanh(V_A x_k); \qquad x_0 \text{ given} \tag{5.80}$$

for three different cases:

1. *System with multiple equilibria:*

$$W_A = \begin{bmatrix} -1.6663 & -0.7588 & 1.1636 \\ -0.0571 & -0.3085 & -0.1793 \\ -0.8565 & -1.0656 & -0.5651 \end{bmatrix}, V_A = \begin{bmatrix} -0.0542 & -0.4616 & -1.1189 \\ -1.1169 & 0.2699 & -0.6723 \\ 0.8167 & -0.1313 & 0.0912 \end{bmatrix}$$

2. *Quasi-periodic behaviour:*

$$W_A = \begin{bmatrix} -1.8020 & 0.4875 & -1.2491 \\ -0.8841 & 0.4576 & 1.4070 \\ -0.2647 & -0.2385 & 0.3027 \end{bmatrix}, V_A = \begin{bmatrix} 1.2887 & -0.2148 & -1.2468 \\ -0.2501 & 0.5668 & 0.2859 \\ -0.7716 & 0.0636 & -0.8706 \end{bmatrix}$$

3. *Chaos:*

$$W_A = \begin{bmatrix} 0.9690 & 0.6967 & 0.2985 \\ -0.7473 & 3.2069 & 0.2840 \\ -2.7960 & 0.5360 & 0.9597 \end{bmatrix}, V_A = \begin{bmatrix} 2.0876 & 0.0173 & 1.1578 \\ 1.5247 & 0.2463 & 0.1619 \\ -0.1953 & -0.8545 & 1.5571 \end{bmatrix}$$

The behaviour of these models is shown in Figure 5.10 for some given initial state vector. In order to globally stabilize these systems we consider the following control laws

1. *Linear static state feedback:*

   Model:

$$x_{k+1} = W_A \tanh(V_A x_k) + B u_k \tag{5.81}$$

   Controller:

$$u_k = K x_k \tag{5.82}$$

   Closed loop system in $NL_q$ form:

$$\begin{bmatrix} x_{k+1} \\ \rho_{k+1} \end{bmatrix} = \Gamma_1 \begin{bmatrix} BK & W_A \\ V_A B K & V_A W_A \end{bmatrix} \begin{bmatrix} x_k \\ \rho_k \end{bmatrix} \tag{5.83}$$

   with $\rho_k = \tanh(V_A x_k)$.

2. *Linear dynamic feedback:*
Model:
$$\left\{ \begin{array}{rcl} x_{k+1} & = & W_A \tanh(V_A x_k) + B u_k \\ y_k & = & C x_k \end{array} \right. \tag{5.84}$$
Controller:
$$\left\{ \begin{array}{rcl} z_{k+1} & = & E z_k + F y_k \\ u_k & = & G z_k \end{array} \right. \tag{5.85}$$
Closed loop system in NL$_q$ form:
$$\begin{bmatrix} x_{k+1} \\ z_{k+1} \\ \rho_{k+1} \end{bmatrix} = \Gamma_1 \begin{bmatrix} 0 & BG & W_A \\ FC & E & 0 \\ 0 & V_A B G & V_A W_A \end{bmatrix} \begin{bmatrix} x_k \\ z_k \\ \rho_k \end{bmatrix} \tag{5.86}$$
with $\rho_k = \tanh(V_A x_k)$.

3. *Nonlinear dynamic feedback:*
Model:
$$\left\{ \begin{array}{rcl} x_{k+1} & = & W_{AB} \tanh(V_A x_k + V_B u_k) \\ y_k & = & C x_k \end{array} \right. \tag{5.87}$$
Controller:
$$\left\{ \begin{array}{rcl} z_{k+1} & = & W_{EF} \tanh(V_E z_k + V_F y_k) \\ u_k & = & G z_k \end{array} \right. \tag{5.88}$$
Closed loop system in NL$_q$ form:
$$\begin{bmatrix} x_{k+1} \\ z_{k+1} \\ \rho_{k+1} \\ \eta_{k+1} \end{bmatrix} = \Gamma_1 \begin{bmatrix} 0 & 0 & W_{AB} & 0 \\ 0 & 0 & 0 & W_{EF} \\ 0 & 0 & V_A W_{AB} & V_B G W_{EF} \\ 0 & 0 & V_F C W_{AB} & V_E W_{EF} \end{bmatrix} \begin{bmatrix} x_k \\ z_k \\ \rho_k \\ \eta_k \end{bmatrix} \tag{5.89}$$
with $\rho_k = \tanh(V_A x_k + V_B G z_k)$, $\eta_k = \tanh(V_E z_k + V_F C x_k)$.

For these problems, it turns out that it is very hard (probably impossible) to find a feasible point to the theorems of diagonal dominance and condition number factor. Nevertheless it is still possible to enforce the closed loop system towards global asymptotic stability, by stating the problem in another way:

- *Diagonal dominance:* One minimizes
$$\min c_\alpha, \text{ s.t. } \|P_{tot} Z_{tot} P_{tot}^{-1}\|_2 < 1. \tag{5.90}$$

- *Condition number factor:* One minimizes
$$\min c_P, \text{ s.t. } \|P_{tot} Z_{tot} P_{tot}^{-1}\|_2 < 1. \tag{5.91}$$

The constraint in (5.90) and (5.91) correspond to local stability at the origin. We show now that making the matrices $Q_i$ in (5.90) as diagonally dominant as possible or making the condition number of the matrices $P_i$ in (5.91) as low as possible, enforces the closed loop system 'towards' global asymptotical stability as well, even if $c_\alpha \beta_P > 1$ or $c_P \beta_P > 1$. So although the conditions of Theorems 5.7 and 5.8 might be too conservative, they turn out to be useful in order to stabilize complex behaviour.

For the system with multiple equilibria a linear static state feedback controller is taken as example. Full state information is assumed and $B = I_3$. Using an SQP method the constrained optimization problem (5.90) is solved with objective $\|PZP^{-1}\|_2 + \|X_Q\|_\infty$ with $Q = P^T PN$, $N$ a diagonal matrix and $\|X_Q\|_\infty$ according to Lemma 5.3. The constraint is chosen as $\|PZP^{-1}\|_2 < 0.95$. The unknown parameter vector consists of the elements of $P$, $N$ and $K$. The gradients of the cost function and the constraint are calculated numerically. For the initial parameter vector $P$ and $N$ were chosen equal to the identity matrix and $K$ according to normal random distribution with zero mean and standard deviation 0.1. A feasible point was found but with $\|X_Q\|_\infty = 1.12 > 1$. The closed loop system was extensively simulated, taking 10000 different initial conditions for $x_0$, random normally distributed with zero mean and variance 3. All these trajectories converged to the origin. Nevertheless it might be possible that there still exist initial states for which the closed loop system would not converge to the origin. Figure 5.11 shows the behaviour of the closed loop system for some random initial state.

For the chaotic system a linear dynamic output feedback controller was taken as example. Full state information was assumed and $B = I_3$, $C = I_3$. A third order controller was proposed. A similar procedure as for the system with multiple equilibria was done with diagonal dominance condition, leading again to similar results. Figure 5.11 shows the behaviour of the closed loop system for some random initial state.

For the quasi-periodic system the neural controller was considered. In this case the condition number factor condition was used. Full state information was assumed and $V_B = I_3$, $C = I_3$. A third order neural controller was proposed with 3 hidden neurons. Again SQP was applied with objective $\|PZP^{-1}\|_2 + \lambda c_P$ with $\lambda = 0.1$ and constraint $\|PZP^{-1}\|_2 < 0.95$. The unknown parameter vector consists of the elements of $P$, $W_{EF}$, $V_E$, $V_F$, $G$. Starting from some given initial parameter vector, a feasible point was immediately found, but with $c_P = 7.60$, such that $c_P \beta_P > 1$. Nevertheless after extensive simulation, we were not able to find one single initial state for which the closed loop system doesn't converge to the origin.

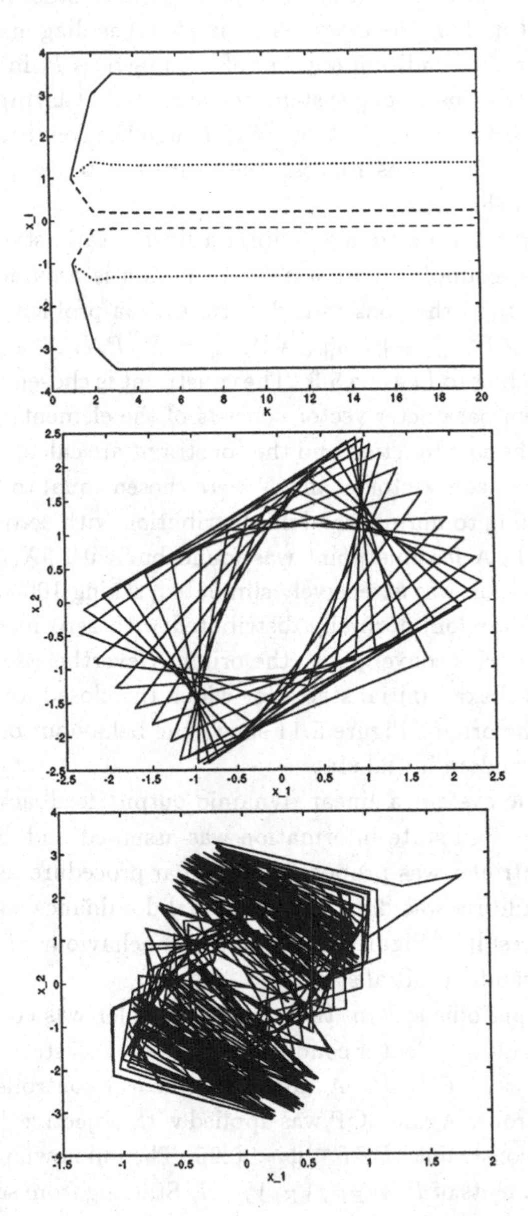


Figure 5.10: ***Neural state space models with different type of behaviour: (Top) system with multiple equilibria, shown is the state vector $x_k$ through time for two different initial states; (Middle) quasi-periodic behaviour, shown is the plane $(x_1, x_2)$; (Bottom) chaos, shown is the plane $(x_1, x_2)$.***

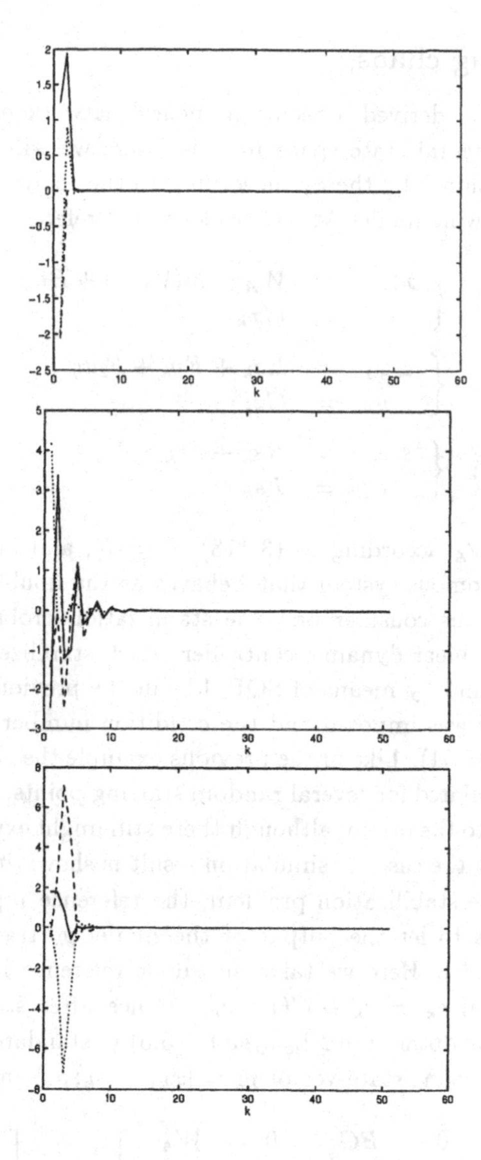


Figure 5.11: *A collection of stabilization problems. (Top) A system with multiple equilibria stabilized by means of linear full static state feedback; (Middle) quasi-periodic behaviour stabilized by means of nonlinear (neural) dynamic feedback; (Bottom) chaos stabilized by means of linear dynamic state feedback. (shown are the state variables $x_k$ through time) $NL_q$ Theorems on diagonal dominance and condition number factor were used in order to enforce the closed loop systems towards global asymptotic stability.*

### 5.8.3 Mastering chaos

In Chapter 3 we have derived a recurrent neural network model for Chua's double scroll using neural state space models. Here we will show how chaos can be mastered within NL$_q$ theory, according to the model-based approach. Let us take the following model $\mathcal{M}$, controller $\mathcal{C}$ and reference model $\mathcal{L}$.

$$\mathcal{M}: \begin{cases} x_{k+1} = W_A \tanh(V_A x_k) + B u_k \\ y_k = C x_k \end{cases}$$

$$\mathcal{C}: \begin{cases} z_{k+1} = E z_k + F y_k + F_2 d_k \\ u_k = G z_k \end{cases}$$

$$\mathcal{L}: \begin{cases} s_{k+1} = R s_k + S r_k \\ d_k = T s_k \end{cases}$$

We take here $W_A$, $V_A$ according to (3.118), $B = I_3$, and full state feedback $C = I_3$. The autonomous system that behaves as the double scroll is shown in Figure 3.11. Let us consider first the stabilization problem, without the reference model. A linear dynamic controller which stabilizes the closed loop system has been found by means of SQP, like in the previous Section. Local asymptotic stability was imposed and the condition number factor was minimized according to (5.91). Like in the previous example the closed loop system was extensively simulated for several random starting points. All observed trajectories converged to the origin, although there still might exist starting points for which this is not the case. A simulation result is shown in Figure 5.12.

After solving the stabilization problem, the reference input $r_k$ was introduced. The goal is to let the output of the model $y_k$ track the output $d_k$ of the reference model. Here we take one single reference input $r_k \in \mathbb{R}$ and take the error signal $e_k = d_k - C(1,:)x_k$. Hence after state augmentation $\rho_k = \tanh(V_A x_k)$, we obtain the NL$_q$ system (5.5) in standard plant form with exogenous input $w_k = r_k$, state vector $p_k = [x_k; z_k; s_k; \rho_k]$ and matrices

$$V_1 = \begin{bmatrix} 0 & BG & 0 & W_A \\ FC & E & F_2T & 0 \\ 0 & 0 & R & 0 \\ 0 & V_ABG & 0 & V_AW_A \end{bmatrix}, B_1 = \begin{bmatrix} 0 \\ 0 \\ S \\ 0 \end{bmatrix} \quad (5.92)$$

$$W_1 = [\ -C(1,:) \quad 0 \quad T \quad 0\ ], D_1 = 0.$$

The following objective was minimized by means of SQP: $\|P_S Z P_S^{-1}\|_2 + 0.1\kappa(P)$ with constraint $\|P_S Z P_S^{-1}\|_2 < 0.95$, where $P_S = \text{blockdiag}\{P, 1\}$ with $P$ a full matrix and $Z = [V_1\ B_1; W_1\ D_1]$. Gradients were calculated numerically. A first order reference model was taken with $R = 0.5$, $S = 0.5$, $T = 1$. The controller is of third order. In Figure 5.12 the output of the model is shown when the reference input $r_k$ is a sine.

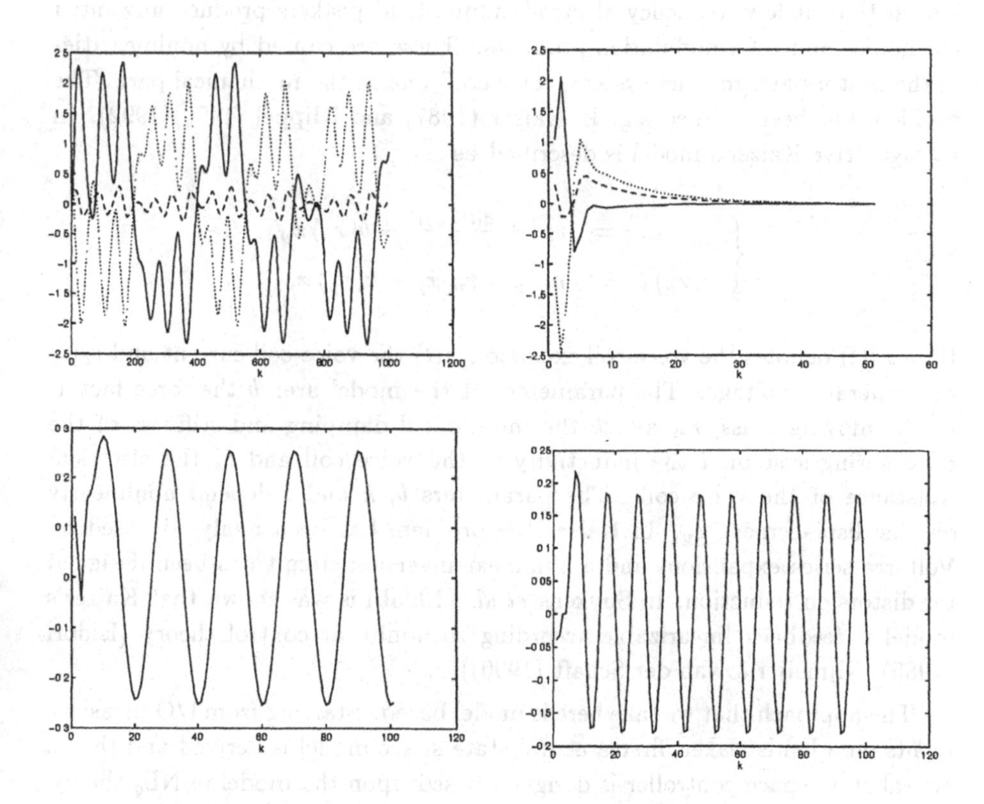

Figure 5.12: *Mastering chaos from a neural state space model for Chua's double scroll. (Top-left) Double scroll behaviour from the neural state space model; (Top-right) Stabilizing the chaotic behaviour towards one single equilibrium point using $NL_q$ theory; (Bottom) Introduction of the reference input in a model reference control scheme. Output of the model that has to track the output of the reference model for sines at different frequencies.*

### 5.8.4 Controlling nonlinear distortion in loudspeakers

In this example we apply NL$_q$ theory to the problem of controlling nonlinear distortion in electrodynamic loudspeakers. The project is in cooperation with Philips (ITCL, Leuven and ASSET, Dendermonde, Belgium). It is well-known that at low frequency electrodynamic loudspeakers produce unwanted harmonics and intermodulation products. These are caused by nonlinearities in the motor part (magnet system/voice coil) and in the mechanical part. The problem has been studied e.g. in Kaizer (1987) and Klippel (1990)(1992). In voltage drive Kaizer's model is described as

$$\left\{ \begin{array}{rcl} v_e & = & r_e\, i + \frac{d(l(x_p)\, i)}{dt} + b(x_p)\, \dot{x}_p \\ b(x_p)\, i & = & m\, \ddot{x}_p + r_m\, \dot{x}_p + k(x_p)\, x_p \end{array} \right.$$

Here $x_p(t)$ denotes the voice-coil excursion, $i(t)$ the voice-coil current and $v_e(t)$ the generator voltage. The parameters of the model are: $b$ the force factor, $m$ the moving mass, $r_m$ and $k$ the mechanical damping and stiffness of the mass-spring system, $l$ the inductivity of the voice coil and $r_e$ the electrical resistance of the voice coil. The parameters $b$, $k$ and $l$ depend nonlinearly on the displacement $x_p$. In Kaizer the problem has been analysed based on Volterra series expansions and a nonlinear inversion circuit has been designed for distorsion reduction. In Suykens *et al.* (1995b) is was shown that Kaizer's model is feedback linearizable according to nonlinear control theory (Isidori (1985), Nijmeijer & van der Schaft (1990)).

The approach that we take here is model based. Starting from I/O measurements on a loudspeaker, first a neural state space model is derived and then a neural state space controller is designed based upon the model in NL$_q$ theory. A Tiny box was selected for the experiments at Philips Dendermonde. I/O data were taken for an input voltage of zero mean white Gaussian noise with standard deviation $7.5V$ fed through a low pass filter with cut off frequency at 205 Hz. As output for the system the pressure was measured at a distance of 0.5m of the speaker. Sampling frequency is 512 Hz. A neural state space model of order 3 was identified with as number of hidden neurons $n_{hx} = 3$, $n_{hy} = 1$ together with nonzero Kalman gain and zero bias vectors. The identified model was acceptable according to the higher order correlation tests.

The following control scheme is then studied:

$$\mathcal{M} : \left\{ \begin{array}{rcl} \hat{x}_{k+1} & = & W_A \tanh(V_A \hat{x}_k + V_B u_k) \\ \hat{y}_k & = & W_C \tanh(V_C \hat{x}_k) \end{array} \right.$$

$$\mathcal{C}: \begin{cases} z_{k+1} &= W_{EF} \tanh(V_E z_k + V_F \hat{x}_k + V_{F_2} d_k) \\ u_k &= G z_k \end{cases}$$

$$\mathcal{L}: \begin{cases} s_{k+1} &= A s_k + B r_k \\ d_k &= C s_k + D r_k \end{cases}$$

where $\mathcal{M}$ is the neural state space model, $\mathcal{C}$ the neural state space controller and $\mathcal{L}$ the linear reference model. In order to compare the distortion of the uncontrolled system with the neural controlled system, the deterministic part of the identified model was selected. The tracking error is defined as $e_k = \hat{y}_k - d_k$. For the NL$_q$ standard plant form $w_k = r_k$ is the exogenous input, $e_k$ the regulated output and $p_k = [\hat{x}_k; z_k; s_k; \rho_k]$ the state of the NL$_q$ with $\rho_k = \tanh(V_C \hat{x}_k)$. The matrices for the NL$_q$ (5.5) are

$$V_2 = \begin{bmatrix} V_A & V_B G & 0 & 0 \\ V_F & V_E & V_{F_2} C & 0 \\ 0 & 0 & A & 0 \\ V_A & V_B G & 0 & 0 \end{bmatrix}, B_2 = \begin{bmatrix} 0 & 0 \\ V_{F_2} D & 0 \\ B & 0 \\ 0 & 0 \end{bmatrix}$$

$$W_1 = I, W_2 = \begin{bmatrix} 0 & 0 & -C & W_C \end{bmatrix}, D_2 = \begin{bmatrix} -D & 0 \end{bmatrix}$$

and $V_1 = \text{blockdiag}\{W_{AB}, W_{EF}, I, V_C W_{AB}\}$, $B_1 = 0$, $D_1 = 0$. Some third order reference model was taken, based on linear system identification from I/O data. A third order neural controller with 3 hidden neurons is chosen. The following optimal performance objective was minimized by means of SQP: $\|P_{tot} Z_{tot} P_{tot}^{-1}\| + \lambda \max\{\kappa(P_1), \kappa(P_2)\}$ under a local stability constraint, where the meaning of $P_{tot}$, $P_1$, $P_2$ and $Z_{tot}$ follows from (5.58) and $\lambda = 0.1$. The unknown parameter vector for the optimization problem consists of the elements of $W_{EF}$, $V_E$, $V_F$, $V_{F_2}$, $G$ and $P_1$, $P_2$. The number of elements of this parameter vector is 358. The optimization procedure was initialized by a random neural controller and matrices $P_1$, $P_2$ equal to the identity matrix. The computational cost for this optimization is typically a few hours on a DEC 5000/240 workstation. In Figure 5.13 and 5.14 we show the frequency response (power spectral density *psd* of Matlab) of the uncontrolled neural state space model together with the neural controlled model on a sine reference input $r_k$ of 20, 40, 60 and 80 Hz with amplitude 40 V (but the plots are scaled in another way, due to the scaling of the input data (divided by 20) for identification purposes). One observes a significant linearization of the neural controlled system with respect to the uncontrolled model, in the sense that the higher harmonics are reduced. Another working approach is to let the neural controller learn to track a set of specific reference inputs in the frequency range of interest by means of the modified dynamic backpropagation algorithm.

Further work remains to be done here on adaptive control strategies (because loudspeaker characteristics vary with time), the selection of good reference models and DSP implementation of the control scheme.

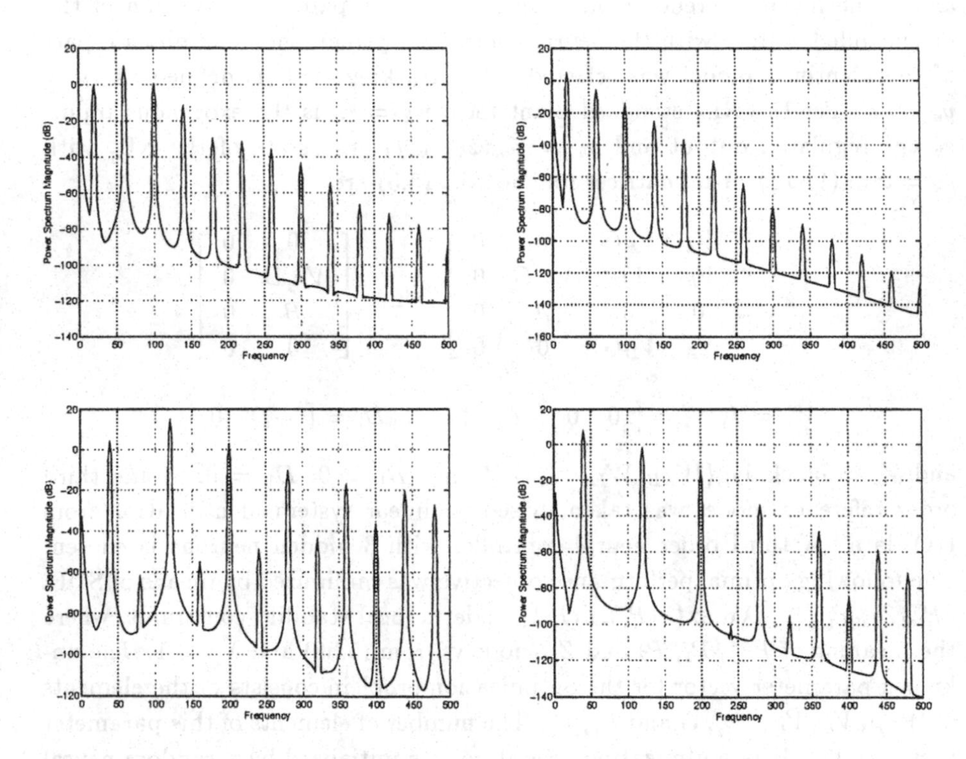


Figure 5.13: *Reducing nonlinear distortion in electrodynamic loudspeakers using a model-based neural control approach within $NL_q$ theory. The frequency response to a sine reference input is shown at 20 Hz (Top) and 40Hz (Bottom). One sees a significant reduction of the higher order harmonics: (Left) frequency response of the identified neural state space model; (Right) frequency response of the neural controlled identified model.*

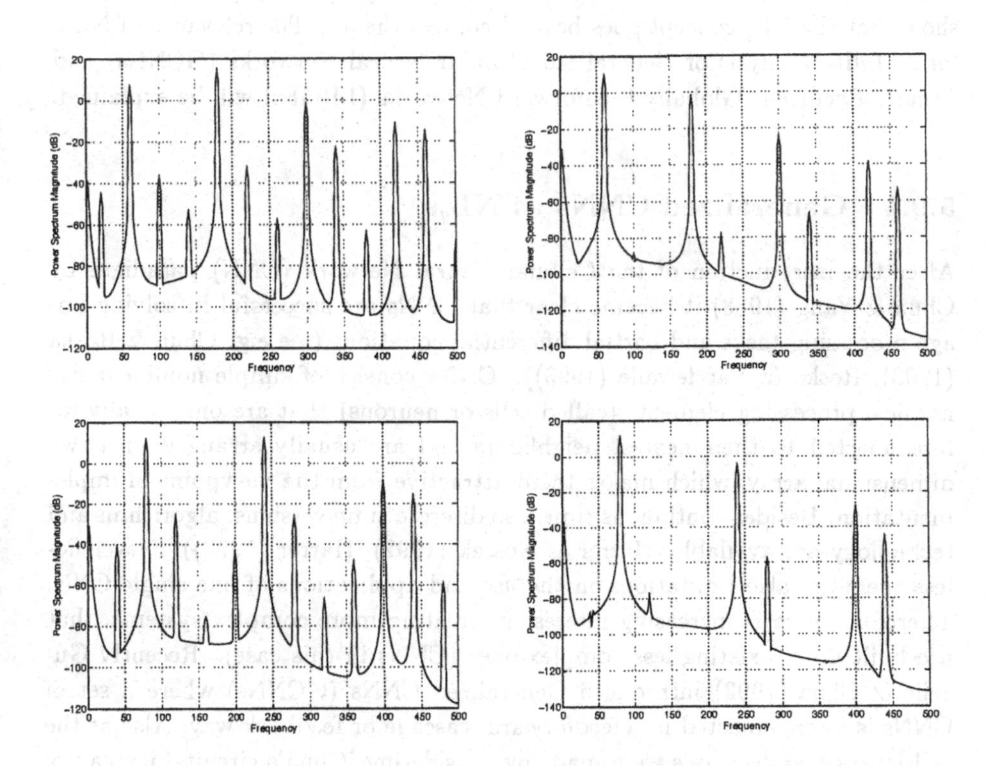


Figure 5.14: *Similar to Figure 5.13, but for (Top) 60 Hz and (Bottom) 80 Hz.*

## 5.9 $NL_q$s beyond control

In the previous Sections of this Chapter we have focused on stability analysis of neural control systems. A key role was played by the $NL_q$ system form. We show that the $NL_q$ concept goes beyond control theory. The relevance of $NL_q$s for stability analysis of Generalized Cellular Neural Networks (GCNNs) and Locally Recurrent Globally Feedforward Networks (LRGFs) will be explained.

### 5.9.1 Generalized CNNs as $NL_q$s

After the introduction of the Cellular Neural Network (CNN) paradigm by Chua & Yang (1988) it became clear that CNNs are successful in solving image processing tasks and partial differential equations (see e.g. Chua & Roska (1993), Roska & Vandewalle (1993)). CNNs consist of simple nonlinear dynamical processing elements (called cells or neurons) that are only locally interconnected to their nearest neighbours and are usually arranged in a two dimensional array, which makes them attractive from the viewpoint of implementation. Besides continuous time also discrete time versions, algorithms and technology are available (Harrer & Nossek (1992), Harrer (1993)). Nevertheless there are also limitations on the use and applications of one single CNN. Therefore there is increasing interest in creating more complex systems, that are built up of existing less complex ones (CNNs in this case). Recently Guzelis & Chua (1993) introduced Generalized CNNs (GCNNs) where a set of CNNs is interconnected in a feedforward, cascade or feedback way. Also at the cell level generalizations were made by considering 'Chua's circuits' instead of 'neurons', leading to cellular 'nonlinear' networks instead of cellular 'neural' networks (Roska & Chua (1993)). Because even simple nonlinear dynamical systems such as Chua's circuit can already exhibit complex behaviour, letting such circuits interact within an array, results in new rich phenomena (see Chua (1994) for an overview). On the other hand stability analysis of interconnected systems becomes more complicated. This motivates the need for general frameworks for the analysis and design of such systems. We illustrate now how $NL_q$ theory may contribute towards this perspective.

Generalized CNNs consist of CNNs, which are interconnected e.g. in a feedforward, feedback or cascade way (see Guzelis & Chua (1993)). They have a certain inter-layer (between the CNNs) and intra-layer (within the CNN) connectivity. Let us consider a GCNN that consists of $L$ CNNs with layer index $l = 1, ..., L$. The dynamics of the $l$-th CNN are assumed to take the discrete time form

$$\begin{cases} x_{k+1}^l &= A^l x_k^l + B^l u_k^l \\ z_k^l &= C^l x_k^l \\ y_k^l &= f(z_k^l) \\ u_k^l &= \sum_i E_i^l y_k^i + \sum_j F_j^l v_k^j + \beta^l, \end{cases} \quad (5.93)$$

with internal state $x_k \in \mathbb{R}^{n_l}$, input to the linear dynamical subcircuit $u_k \in \mathbb{R}^{m_l}$, output $y_k \in \mathbb{R}^{p_l}$ and external input $v_k \in \mathbb{R}^{q_l}$. The upper index of the matrices and vectors is the layer index. The notation $E_i^l$, $F_j^l$ means that the $l$-th CNN is connected to the output of the $i$-th CNN and the $j$-th external input respectively. The discrete time index is $k$. The nonlinearity $f(.)$ is taken elementwise when applied to a vector. If $f(.)$ is a static nonlinearity belonging to sector $[0, 1]$ we will use the notation $\sigma(.)$ instead of $f(.)$ in the sequel.

Alternatively the dynamics (5.93) is written as

$$\begin{cases} x_{k+1}^l &= A^l x_k^l + B^l E_l^l f(C^l x_k^l) + B^l \sum_{i \neq l} E_i^l f(C^i x_k^i) + B^l \sum_j F_j^l v_k^j + B^l \beta^l \\ y_k^l &= f(C^l x_k^l). \end{cases} \quad (5.94)$$

The dynamics (5.93) is formulated at the CNN level and not at the cell level as was done by Guzelis & Chua. We make abstraction here of the intra-layer connectivity. The local interconnection between the cells is then revealed by the sparseness of certain matrices in (5.93).

We now illustrate by examples how discrete time CNNs and GCNNs can be represented as $NL_q$s (see Suykens & Vandewalle (1995)).

**Example 5.4.** Consider a CNN of the form

$$x_{k+1} = Ax_k + B\sigma(Cx_k) + Fv_k + \beta. \quad (5.95)$$

Using the state augmentation

$$\xi_k = \sigma(Cx_k)$$

one obtains the state space description

$$\begin{cases} x_{k+1} &= Ax_k + B\xi_k + Fv_k + \beta \\ \xi_{k+1} &= \sigma(CAx_k + CB\xi_k + CFv_k + C\beta) \end{cases}$$

Defining $p_k = [x_k; \xi_k]$ and $w_k = [v_k; 1]$ one has the $NL_1$ system

$$p_{k+1} = \Gamma_1(V_1 p_k + B_1 w_k) \quad (5.96)$$

with matrices

$$V_1 = \begin{bmatrix} A & B \\ CA & CB \end{bmatrix}, B_1 = \begin{bmatrix} F & \beta \\ CF & C\beta \end{bmatrix}$$

and $\Gamma_1 = \text{diag}\{I, \Gamma(x_k, \xi_k, v_k)\}$ with $\|\Gamma_1\| \leq 1$.

**Example 5.5.** Consider a Generalized CNN according to Figure 5.15 which consists of 6 CNNs, containing feedforward, feedback and cascade interconnections. The system is described as

$$\left\{ \begin{array}{lcl} x_{k+1}^1 & = & A^1 x_k^1 + B^1 \sigma(C^1 x_k^1) + E_5^1 \sigma(C^5 x_k^5) + F^1 v_k + \beta^1 \\ x_{k+1}^2 & = & A^2 x_k^2 + B^2 \sigma(C^2 x_k^2) + E_1^2 \sigma(C^1 x_k^1) + E_6^2 \sigma(C^6 x_k^6) + \beta^2 \\ x_{k+1}^3 & = & A^3 x_k^3 + B^3 \sigma(C^3 x_k^3) + E_2^3 \sigma(C^2 x_k^2) + \beta^3 \\ x_{k+1}^4 & = & A^4 x_k^4 + B^4 \sigma(C^4 x_k^4) + E_2^4 \sigma(C^2 x_k^2) + E_3^4 \sigma(C^3 x_k^3) + \beta^4 \\ x_{k+1}^5 & = & A^5 x_k^5 + B^5 \sigma(C^5 x_k^5) + E_4^5 \sigma(C^4 x_k^4) + \beta^5 \\ x_{k+1}^6 & = & A^6 x_k^6 + B^6 \sigma(C^6 x_k^6) + E_4^6 \sigma(C^4 x_k^4) + \beta^6 \end{array} \right. \quad (5.97)$$

with external input $v_k$. Applying the state augmentation

$$\xi_k^i = \sigma(C^i x_k^i), \quad i = 1, 2, ..., 6$$

and defining

$$p_k = [x_k^1; x_k^2; x_k^3; x_k^4; x_k^5; x_k^6; \xi_k^1; \xi_k^2; \xi_k^3; \xi_k^4; \xi_k^5; \xi_k^6]$$

and $w_k = [v_k; 1]$, one obtains again an $\text{NL}_q$ system with $q = 1$ and matrices

$$V_1 = \begin{bmatrix} A^1 & & & & & & B^1 & & & & E_5^1 & \\ & A^2 & & & & & E_1^2 & B^2 & & & & E_6^2 \\ & & A^3 & & & & & E_2^3 & B^3 & & & \\ & & & A^4 & & & & E_2^4 & E_3^4 & B^4 & & \\ & & & & A^5 & & & & & E_4^5 & B^5 & \\ & & & & & A^6 & & & & E_4^6 & & B^6 \\ C^1A^1 & & & & & & C^1B^1 & & & & C^1E_5^1 & \\ & C^2A^2 & & & & & C^2E_1^2 & C^2B^2 & & & & C^2E_6^2 \\ & & C^3A^3 & & & & & C^3E_2^3 & C^3B^3 & & & \\ & & & C^4A^4 & & & & C^4E_2^4 & C^4E_3^4 & C^4B^4 & & \\ & & & & C^5A^5 & & & & & C^5E_4^5 & C^5B^5 & \\ & & & & & C^6A^6 & & & & C^6E_4^6 & & C^6B^6 \end{bmatrix}$$

$$B_1 = \begin{bmatrix} \beta^1 & \beta^2 & \beta^3 & \beta^4 & \beta^5 & \beta^6 & C^1\beta^1 & C^2\beta^2 & C^3\beta^3 & C^4\beta^4 & C^5\beta^5 & C^6\beta^6 \\ F^1 & 0 & 0 & 0 & 0 & 0 & C^1F^1 & 0 & 0 & 0 & 0 & 0 \end{bmatrix}^T. \quad (5.98)$$

Hence although the interconnectivity of this GCCN is rather complex, the overall system description has still low complexity ($q = 1$) if we consider the $q$

value of the $NL_q$ to be a measure of complexity for the overall system. This is due to the fact that the CNNs contain a linear dynamical subcircuit. Neural state space models and neural control systems may lead to $q > 1$ values because of their multilayer feature. In order to create CNNs and GCNNs with $q > 1$, let us modify the nonlinearity $f(.)$ in (5.93). For CNNs like (5.95) the nonlinearity $f(.)$ usually satisfies a sector condition. However it has been shown by Guzelis & Chua that Chua's circuit can be written in a form like (5.93) (but in continuous time and on the cell level). The nonlinearity of the nonlinear resistor in Chua's circuit plays then the role of the nonlinearity $f(.)$ in the continuous time version to (5.93). Hence instead of taking $y_k^l = \sigma(C^l z_k^l)$ in (5.93) a useful extension is to take

$$y_k^l = \sigma(W^l \sigma(V^l x_k^l + \delta^l))$$

where $W^l \sigma(V^l x_k^l + \delta^l)$ is a multilayer perceptron with interconnection matrices $W^l$, $V^l$ and bias vector $\delta^l$, which is able to represent any continuous nonlinear mapping (including the nonlinearity used in Chua's circuit), provided there are 'enough' hidden neurons (see Chapter 2). This leads to the type of recurrent neural network proposed in the following example.

**Example 5.6.** The use of a multilayer perceptron for $f(.)$ in (5.93) gives

$$x_{k+1} = Ax_k + B\sigma(W\sigma(Vx_k + \delta)) + Fv_k + \beta. \tag{5.99}$$

Applying the state augmentation

$$\xi_k = \sigma(W\sigma(Vx_k + \delta))$$

one obtains the state space description

$$\begin{cases} x_{k+1} &= Ax_k + B\xi_k + Fv_k + \beta \\ \xi_{k+1} &= \sigma(W\sigma(VAx_k + VB\xi_k + VFv_k + V\beta + \delta)). \end{cases}$$

Defining $p_k = [x_k; \xi_k]$, $w_k = [v_k; 1]$ this yields an $NL_q$ system with $q = 2$ and

$$V_1 = \begin{bmatrix} I & 0 \\ 0 & W \end{bmatrix}, V_2 = \begin{bmatrix} A & B \\ VA & VB \end{bmatrix}, B_2 = \begin{bmatrix} F & \beta \\ VF & V\beta \end{bmatrix}, B_1 = 0. \tag{5.100}$$

A similar GCCN like in Example 5.5 could be considered, which consists of CNNs of the form (5.99). This would also lead to an $NL_2$.

Hence the GCNNs are representable as $NL_q$s. The sufficient global asymptotic and I/O stability criteria, derived in this Chapter, are then directly applicable to GCNNs. Checking these criteria involves the solution to a convex optimization problem.

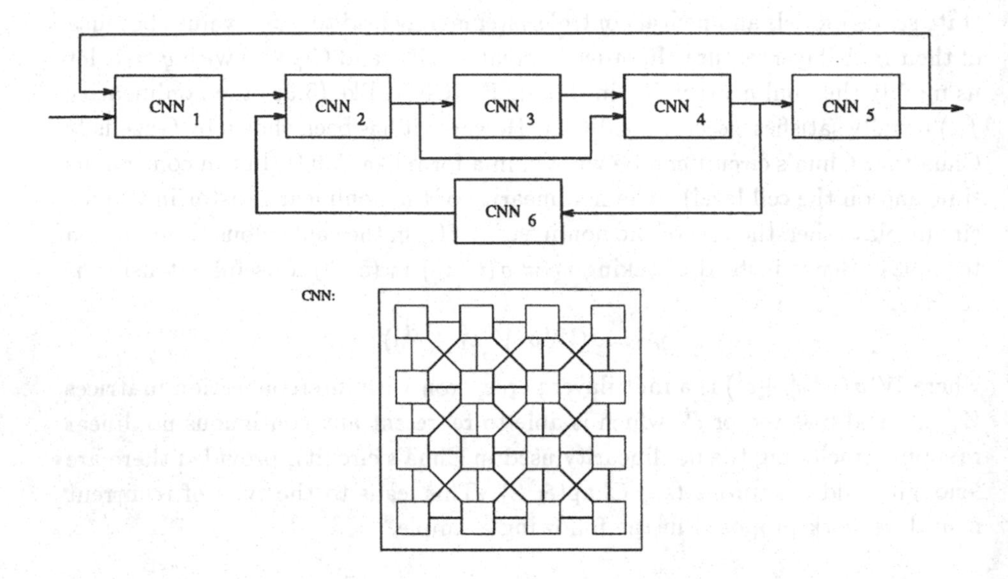


Figure 5.15: *Generalized Cellular Neural Networks (GCNNs): (Top) Example of a GCNN that consists of 6 CNNs that are interconnected in a feedforward, cascade and feedback way. Existing GCNNs can be represented as $NL_q$ systems with $q = 1$. (Bottom) Example of two-dimensional CNN. The squares are dynamical cells (neurons) that are locally interconnected to their nearest neighbours.*

## 5.9.2 LRGF networks as $NL_q$s

In Tsoi & Back (1994) a recurrent neural network architecture, called LRGF (Locally Recurrent Globally Feedforward network), has been proposed, that aims at unifying several existing recurrent neural network models. Starting from the McCulloch-Pitts model, that is a simple static nonlinear model for a neuron, many other architectures were proposed in the past with local synapse feedback, local activation feedback and local output feedback, time delayed neural networks etc. (see Figure 5.16), e.g. by Frasconi-Gori-Soda, De Vries-Principe and Poddar-Unnikrishnan. The architecture of Tsoi & Back, which is shown in Figure 5.16, includes most of the other architectures.

We show now how this LRGF network can be represented as an $NL_q$ system. Assuming the transfer functions $G_i(z)$ $(i = 1, .., n)$ (which may have both poles and zeros) have a state space realization $(A_i, B_i, C_i)$ the LRGF can be described in state space form as

$$\begin{cases} \xi_{k+1}^{(i)} &= A^{(i)}\xi_k^{(i)} + B^{(i)}u_k^{(i)}, \qquad i = 1, ..., n-1 \\ z_k^{(i)} &= C^{(i)}\xi_k^{(i)} \\ & \\ \xi_{k+1}^{(n)} &= A^{(n)}\xi_k^{(n)} + B^{(n)}f(\sum_{j=1}^{n} z_k^{(j)}) \\ z_k^{(n)} &= C^{(n)}\xi_k^{(n)} \\ y_k &= f(\sum_{j=1}^{n} z_k^{(j)}). \end{cases}$$

Applying the state augmentation $\eta_k = f(\sum_{j=1}^{n} z_k^{(j)})$ an NL$_1$ system is obtained with state vector $p_k = [\xi_k^{(1)}; \xi_k^{(2)}; ...; \xi_k^{(n-1)}; \xi_k^{(n)}; \eta_k]$ and input vector $w_k = [u_k^{(1)}; ...; u_k^{(n-1)}]$ in (5.5) and matrices

$$\mathrm{V}_1 = \begin{bmatrix} A^{(1)} & & & & 0 & 0 \\ & A^{(2)} & & & & \\ & & \ddots & & & \vdots \\ & & & A^{(n-1)} & 0 & 0 \\ 0 & 0 & \dots & 0 & A^{(n)} & B^{(n)} \\ C^{(1)}A^{(1)} & C^{(2)}A^{(2)} & \dots & C^{(n-1)}A^{(n-1)} & C^{(n)}A^{(n)} & C^{(n)}B^{(n)} \end{bmatrix}$$

$$\mathrm{B}_1 = \begin{bmatrix} B^{(1)} & & & \\ & B^{(2)} & & \\ & & \ddots & \\ & & & B^{(n-1)} \\ 0 & 0 & \dots & 0 \\ C^{(1)}B^{(1)} & C^{(2)}B^{(2)} & \dots & C^{(n-1)}B^{(n-1)} \end{bmatrix}$$

We assume here that $f(.)$ is a static nonlinearity belonging to sector $[0, 1]$.

As a result properties of the LRGF architecture, such as global asymptotic stability and I/O stability with finite $L_2$-gain, can be studied as a function of the choice of the transfer functions $G_i(z)$ by means of the stability and dissipativity Theorems of NL$_q$ theory.

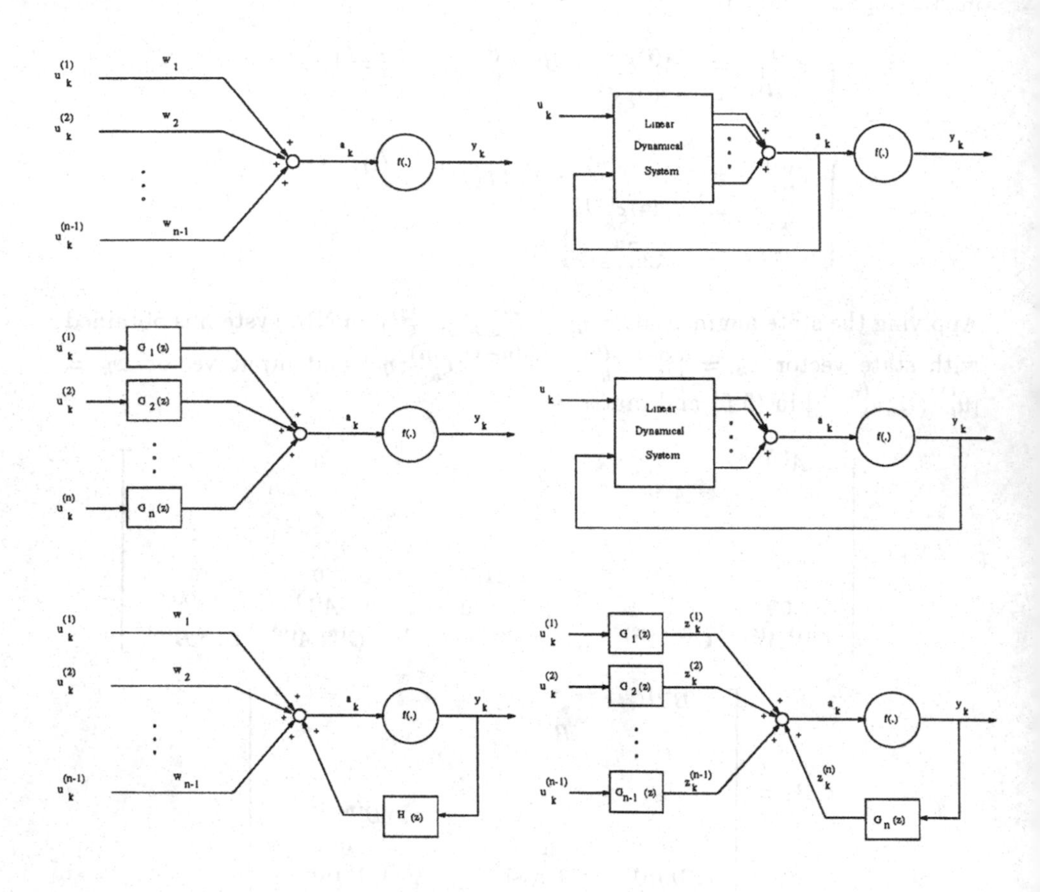


Figure 5.16: *Some types of recurrent neural networks, according to Tsoi & Back (1994): (Top-left) McCulloch-Pitts neuron or Lapedes-Farber architecture for a neuron, which is static nonlinear; (Top-right) Local activation feedback architecture; (Middle-left) Local synapse feedback architecture; (Middle-right) Local output feedback architecture; (Bottom-left) Generalized Frasconi-Gori-Soda architecture. In the Frasconi-Gori-Soda architecture $H(z)$ is a tapped delay line; (Bottom-right) General locally recurrent globally feedforward network architecture (LRGF) of Tsoi & Back. The network includes local synapse feedback as well as local output feedback.*

## 5.10 Conclusion

The new concept of $\mathrm{NL}_q$ systems plays a unifying role for many problems arising in classical, modern and neural control theory and neural networks. The following systems were proven to be specific examples of $\mathrm{NL}_q$s: all neural state space control problems $\Xi_j^i$s, the discrete time Lur'e problem, LFTs with real diagonal $\Delta$ block, Hopfield nets, generalized CNNs, locally recurrent globally feedforward neural nets and digital filters with overflow characteristic. A common aspect is that static nonlinearities belonging to a sector $[0, K]$ are involved. Some $\mathrm{NL}_q$ Theorems reduce for $q = 1$ to well-known results in modern control theory. Because of this unifying nature, theorems that are proven for $\mathrm{NL}_q$s may have a simultaneous impact upon the different fields. Here such theorems were derived for global asymptotic stability and dissipativity with finite $L_2$-gain of $\mathrm{NL}_q$s and the links were revealed with modern control theory ($\mathrm{H}_\infty$ control theory and $\mu$-theory). However open problems exist such as e.g. necessary and sufficient conditions for global asymptotic stability (the conditions derived in this book are only sufficient) and conditions for asymptotic stability (which allows the existence of more than one equilibrium point). Another feature is that all systems considered in the $\mathrm{NL}_q$ framework are in state space form: this made it easy to define state augmentations in order to convert a given system into an $\mathrm{NL}_q$. Vector norms as Lyapunov functions and properties of induced norms served as tools to derive the main results.

For control design efficient algorithms of convex optimization can be used for checking stability and dissipativity of the learned controller (analysis problem). In the synthesis problem a modified dynamic backpropagation algorithm was proposed such that global stability or dissipativity for the closed loop system can be ensured. Linear matrix inequalities play a key role in these problems. Even when it's impossible to find a feasible point to the $\mathrm{NL}_q$ theorems, their principles are still valuable in order to enforce complex nonlinear behaviour towards global asymptotic stability. This is done by imposing local stability and making certain matrices as diagonally dominant as possible or minimizing a certain condition number factor. The latter was e.g. illustrated on mastering chaos in a neural state space model emulator for Chua's double scroll attractor.

# Chapter 6

# General conclusions and future work

In this book we discussed the use of artificial neural networks for modelling and control of nonlinear systems in a systemtheoretical context. After a short introduction on neural information processing systems in Chapter 1, we have reviewed basic neural network architectures and their learning rules in Chapter 2, for feedforward as well as recurrent networks. In Chapter 3 we have treated the problem of nonlinear system identification using neural networks. Existing models such as NARX and NARMAX were discussed and neural state space models are introduced. Off- and on-line learning algorithms are presented. An interpretation of neural network models as uncertain linear systems, representable as linear fractional transformations, has been given. Examples on nonlinear system identification of a simulated nonlinear system with hysteresis, a glass furnace with real data and chaotic systems, show the effectiveness of neural state space models. In Chapter 4 a short overview of neural control strategies is given. Neural optimal control has been discussed in more detail. The stabilization problem and tracking problem are discussed. Furthermore it is shown how results from linear control theory can be used as constraint on the neural control design, in order to achieve local stability at a target point. The latter method has been successfully applied to the problems of swinging up an inverted and double inverted pendulum. In Chapter 5 we introduced a neural state space model and control framework with stability criteria. Closed loop systems were transformed into $NL_q$ system forms. Sufficient conditions for global asymptotic stability and I/O stability with finite $L_2$-gain are derived. Links with $H_\infty$ control and $\mu$ theory are revealed. The criteria are formulated as linear matrix inequalities. $NL_q$ theory has been applied to the control of several types of nonlinear behaviour, including chaos and to the real life example of controlling nonlinear distortion in electrodynamic loudspeakers. Furthermore, several types of recurrent neural networks are represented as $NL_q$s, such as generalized cellular neural networks, multilayer Hopfield networks and locally recurrent globally feedforward networks.

Hence, the area of neural control, which started somewhat ten years ago with the broomstick balancing experiment, matured in recent years. Now theory is available on neural optimal control, reinforcement learning, neural adaptive control, feedback linearization, internal model control and predictive control using neural nets etc. However one important link with modern control theory and robust control was still missing. This link is precisely revealed in this book by means of $NL_q$ theory. A model-based neural control framework has been developed in standard plant form according to modern control theory. It turns out that $NL_q$ theory is unifying in nature. While linear models and controllers are studied in modern control theory, $NL_q$ systems are an extension towards layered nonlinear systems, such as dynamical systems that contain multilayer neural nets. Are we realizing here another part of the Cybernetics dream (Wiener (1961)), where control, information and neural science were once regarded under the same banner ? Maybe, because we find how the fields of neural networks, systems and control are converging to each other, at least with respect to stability criteria from $NL_q$ theory. Concepts like dissipativity and linear matrix inequalities play a central role. In fact neural networks have served as a catalyst for extending modern linear control theory towards nonlinear $NL_q$ theory.

A nice property of neural network models is that they are very general in the sense that they can model a large class of nonlinear phenomena and on the other hand remain mathematically tractable. In $NL_q$ theory top-down neural control design is possible. Starting from input/output measurements on a nonlinear system, it is possible to derive a neural state space model and to design neural state space controllers based upon the identified models. We have shown by examples that many types of behaviour can be mastered within $NL_q$ theory. Hence for black box approaches the next logical step to go from linear towards nonlinear modelling and control might be to choose for neural networks and $NL_q$ theory.

With respect to future work and possible developments the following remarks can be some source of inspiration. $NL_q$ theory was developed in a discrete time setting and in a state space approach. Continuous time systems and

I/O models are still to be studied. Also frequency domain approaches need to be developed. A complete understanding of the behaviour of $NL_q$ systems would be important and would have an immediate impact on the theory of neural networks, systems and control. In this book we have derived sufficient global asymptotic stability criteria. Finding asymptotic stability criteria and necessary and sufficient conditions for global asymptotic stability (or at least trying to find the sharpest criteria) is important. For linear systems we know that the class of all linear stabilizing controllers is characterized by the Youla parametrization; finding a parametrization for the class of all stabilizing controllers for $NL_q$ systems would be interesting also. Quadratically convergent learning algorithms for learning within $NL_q$ theory are needed, which highly depends on the status of the field of non-convex nondifferentiable optimization. Also adaptive methods for $NL_q$ theory can be investigated and other norms than those in $l_2$ can be considered.

We already mentioned the possibility for top-down neural control design. One might think here of 'neural control compilers' and CAD in order to automatically identify models and controllers based upon the identified model. With respect to identification further work remains to be done on the choice of the complexity of the models and specialized learning algorithms for recurrent networks.

With respect to generalized cellular neural networks, it has been shown how they can be represented as $NL_q$s. On the other hand we have seen that Chua's double scroll can be identified by means of neural state space models, and controlled within $NL_q$ theory. While CNNs are nowadays mainly applied in image processing and solving PDEs, $NL_q$ theory may lead to new applications for generalized CNNs to control, including mastering chaos.

I/O models are still to be studied. Moreover, cross-domain approaches need to be investigated. A more thorough understanding of the behaviour of $NL_q$ systems would be important and would have an immediate impact on the theory of neural state space models and control. In this book we have derived sufficient global asymptotic stability criteria. Finding also local stability criteria and necessary and sufficient conditions for global asymptotic stability (or at least trying to find these for simple examples) is important. For linear systems we know that the class of all linear stabilizing controllers is characterized by the Youla parametrization. A characterization of the class of all stabilizing controllers for $NL_q$ systems would be interesting also. Generally, convergent learning algorithms for learning within $NL_q$ theory are needed, which mainly depends on the status of the field of non-convex and nondifferentiable optimization. Also adaptive methods for $NL_q$ theory can be investigated and other forms than those in [illegible] considered.

We already mentioned the possibility for top-down neural control design. One might think here of 'neural control compilers' and CAD in order to automatically identify models and controllers based upon the identified model. With respect to identification further work remains to be done on the choice of the complexity of the models and specialized learning algorithms for recurrent networks.

With respect to generalized cellular neural networks, it has been shown how they can be represented as $NL_q$s. On the other hand we have seen that Chua's double scroll can be identified by means of neural state space models and controlled within $NL_q$ theory. While CNNs are nowadays mainly applied in image processing and solving PDEs, $NL_q$ theory may lead to new applications for generalized CNNs for control, including mastering chaos.

# Appendix A

# Generation of $n$-double scrolls

The last 10 years Chua's circuit has become a paradigm for chaos in electrical circuits (Chua (1994), Madan (1993)). The circuit shows several types of behaviour between order and chaos ranging from sinks through periodic behaviour to the well-known double scroll attractor (see e.g. Chua *et al.* (1986)). Several types of generalizations to Chua's circuit have been made. An historical overview, generalizations and recent applications are discussed in Chua (1994).

The type of generalization that has been made in Suykens & Vandewalle (1991)(1993a,d) is by modifying the nonlinear resistor of the circuit, in order to construct more complicated attractors that are called $n$-double scrolls ($n = 1, 2, 3, 4, ...$). This is done by introducing additional break points in the characteristic of the nonlinear resistor. The new circuit is a generalization in the sense that the 1-double scroll corresponds to the classical double scroll. Depending on a bifurcation parameter of the circuit one goes from multiple sink to multiple scroll portraits, like one bifurcates from a sink to a double scroll in the original circuit. Computer simulations of the 2-double scroll were recently confirmed by a real electrical circuit, that has been implemented as a generalized cellular neural network by Arena *et al.* (1994). In Chua (1994) a list of applications of the Chua's circuit is given. Examples include e.g. secure communication and music. Spiral waves have been generated by means of an array of Chua's oscillators (Perez-Munuzuri *et al.* (1993)). It can be expected that replacing the original Chua's circuit by an $n$-double scroll circuit will be meaningful in the context of these applications and may lead to new and rich phenomena. It may also contribute towards modelling of chaotic dynamics in neural pattern recognition (Yao & Freeman (1990), Freeman (1993)).

This Appendix is organized as follows. In Section A.1 the generalized Chua's circuit is discussed. In Section A.2 $n$-double scrolls are presented.

## A.1 A generalization of Chua's circuit

The electrical circuit of Figure A.1 with circuit dynamics described as

$$\begin{cases} C_1 \frac{dv_{C_1}}{dt} &= G(v_{C_2} - v_{C_1}) - g(v_{C_1}) \\ C_2 \frac{dv_{C_2}}{dt} &= G(v_{C_1} - v_{C_2}) + i_L \\ L \frac{di_L}{dt} &= -v_{C_2} \end{cases} \tag{A.1}$$

is known as Chua's circuit. Here $v_{C_1}$,$v_{C_2}$,$i_L$ denote respectively the voltage across $C_1$ and $C_2$ and the current through $L$ and $g(v_{C_1})$ is the piecewise-linear function of Figure A.1 consisting of two breakpoints

$$g(v_{C_1}) = m_0 v_{C_1} + 0.5(m_1 - m_0)|v_{C_1} + B_p| + 0.5(m_0 - m_1)|v_{C_1} - B_p| \tag{A.2}$$

By setting the parameters $1/C_1 = 9$, $1/C_2 = 1$, $1/L = 7$, $G = 0.7$, $m_0 = -0.5$, $m_1 = -0.8$, $B_p = 1$ chaotic behaviour is obtained (double scroll attractor). (A.1) can be written in the form

$$\dot{\mathbf{x}} = A(\mathbf{x})\mathbf{x} \tag{A.3}$$

as

$$\begin{bmatrix} \dot{x} \\ \dot{y} \\ \dot{z} \end{bmatrix} = \begin{bmatrix} -a - k(x) & a & 0 \\ b & -b & 1 \\ 0 & -c & 0 \end{bmatrix} \begin{bmatrix} x \\ y \\ z \end{bmatrix} \tag{A.4}$$

or $\dot{\mathbf{x}} = \mathbf{f}(\mathbf{x})$ where $\mathbf{x}(t) = [x(t)\ y(t)\ z(t)]^T$, $\mathbf{f} : \mathbb{R}^3 \mapsto \mathbb{R}^3$, $x = v_{C_1}$, $y = v_{C_2}$, $z = i_L$, $a = G/C_1$, $b = G/C_2$, $c = 1/L$, $k(x) = (1/C_1)g(x)/x$ $(x \neq 0)$ and $C_2 = 1$.

A new circuit is defined now in terms of the state space description (A.4) where the nonlinear function $k(x)$ will be *parametrized* as $k_q(x)$ $(q \in \mathbb{N}, q \geq 1)$. The nonlinear function $g(x)$ is adapted accordingly into $g_q(x) = C_1 k_q(x) x$. The circuit is described by

$$\begin{cases} \dot{x} &= (-a - k_q(x))x + ay \\ \dot{y} &= bx - by + z \\ \dot{z} &= -cy \end{cases} \tag{A.5}$$

with the following algorithmic description for $k_q(x)$

$$
\begin{array}{ll}
\text{if } 0 < |x| \le \delta_1 : & k_q(x) = \alpha_1 \\
\text{for } i = 2 \textit{ to } q & \\
\quad j = 2 + 3(i-2) & \\
\quad \text{if } \delta_{j-1} < |x| \le \delta_j : & k_q(x) = \alpha_2 \frac{|x| - \delta_{j-1}}{\delta_j - \delta_{j-1}} + \alpha_1 \\
\quad \text{if } \delta_j < |x| \le \delta_{j+1} : & k_q(x) = -\alpha_2 \frac{|x| - \delta_j}{\delta_{j+1} - \delta_j} + \alpha_1 + \alpha_2 \\
\quad \text{if } \delta_{j+1} < |x| \le \delta_{j+2} : & k_q(x) = \alpha_1 \\
\text{end} & \\
\text{if } \delta_{4+3(q-2)} < |x| : & k_q(x) = (\alpha_3/|x|)(\beta_1|x| + \beta_2||x| + \delta_{4+3(q-2)}| + \ldots \\
 & \beta_3||x| - \delta_{4+3(q-2)}|)
\end{array}
\tag{A.6}
$$

and by definition

$$
\Delta_q = [\delta_1 \; \delta_2 \; \ldots \; \delta_{4+3(q-2)}]^T . \tag{A.7}
$$

The graphical description of some $k_q(x)$ and $g_q(x)$ is given in Figure A.2. It is easily verified that for $q = 1$ the circuit (A.5)(A.6) corresponds to Chua's circuit (A.4) with $k_1(x) = k(x)$. We will only consider bifurcations here with respect to the parameter $a$ and take $b = 0.7$, $c = 7$, $\alpha_1 = -8a/7$, $\alpha_2 = 2a/7$, $\alpha_3 = a/0.7$, $\beta_1 = -0.5$, $\beta_2 = -0.15$, $\beta_3 = 0.15$.

A local stability analysis shows that there are $4q - 1$ equilibrium points: $\mathbf{eq}_0 = \mathbf{0}$ and $\mathbf{eq}^{\pm}_{j,j-1}$, $\mathbf{eq}^{\pm}_{j+1,j}$, $\mathbf{eq}^{\pm}_{4+3(q-2)}$ with $j = 2 + 3(i-2)$ and $i = 2, \ldots, q$ with $q \ge 1$. The latter appear in pairs ($\pm$ refers to the sign of the $x$ component) with $x$ components satisfying

$$
\begin{array}{ll}
\text{case } \delta_{j-1} < |x| \le \delta_j : & x_{\mathbf{eq}^{\pm}_{j,j-1}} = \pm 0.5(\delta_j + \delta_{j-1}) \\
\text{case } \delta_j < |x| \le \delta_{j+1} : & x_{\mathbf{eq}^{\pm}_{j+1,j}} = \pm 0.5(\delta_{j+1} + \delta_j) \\
\text{case } \delta_{4+3(q-2)} < |x| : & \alpha_3(\beta_1|x_{\mathbf{eq}^{\pm}_{4+3(q-2)}}| + \beta_2||x_{\mathbf{eq}^{\pm}_{4+3(q-2)}}| + \delta_{4+3(q-2)}| + \ldots \\
 & \beta_3||x_{\mathbf{eq}^{\pm}_{4+3(q-2)}}| - \delta_{4+3(q-2)}|) = -a|x_{\mathbf{eq}^{\pm}_{4+3(q-2)}}|.
\end{array}
\tag{A.8}
$$

Evaluated at the equilibrium points this gives

$$
J(\mathbf{eq}_0) = \begin{bmatrix} -a - \alpha_1 & a & 0 \\ b & -b & 1 \\ 0 & -c & 0 \end{bmatrix}
$$

$$J(\mathbf{eq}^{\pm}_{j,j-1}) = \begin{bmatrix} -(a/7)\gamma_{j,j-1} & a & 0 \\ b & -b & 1 \\ 0 & -c & 0 \end{bmatrix}$$

$$J(\mathbf{eq}^{\pm}_{j+1,j}) = \begin{bmatrix} (a/7)\phi_{j+1,j} & a & 0 \\ b & -b & 1 \\ 0 & -c & 0 \end{bmatrix}$$

$$J(\mathbf{eq}^{\pm}_{4+3(q-2)}) = \begin{bmatrix} -a-\alpha_3(\beta_1+\beta_2+\beta_3) & a & 0 \\ b & -b & 1 \\ 0 & -c & 0 \end{bmatrix}$$

with $\gamma_{j,j-1} = \frac{\delta_j+\delta_{j-1}}{\delta_j-\delta_{j-1}}$ and $\phi_{j+1,j} = \frac{\delta_{j+1}+\delta_j}{\delta_{j+1}-\delta_j}$ for $j = 2+3(i-2)$ and $i = 2, ..., q$.

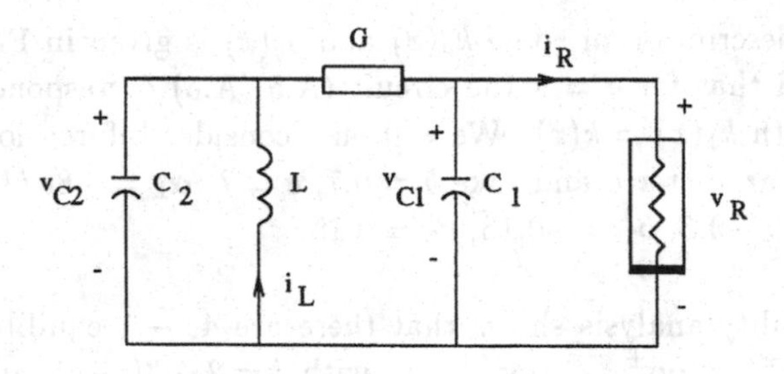


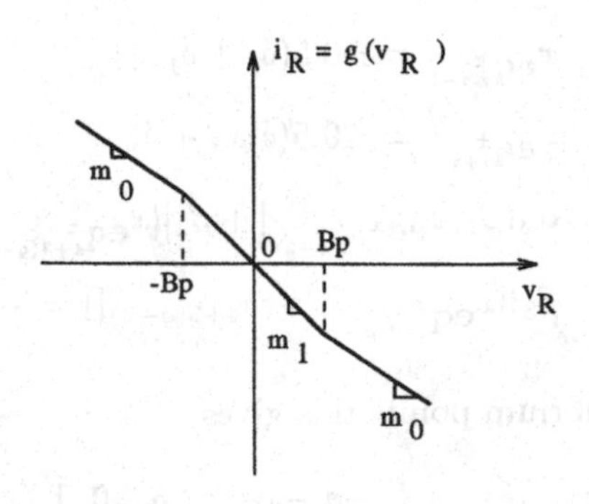


Figure A.1: *Chua's circuit for generation of the double scroll attractor.*

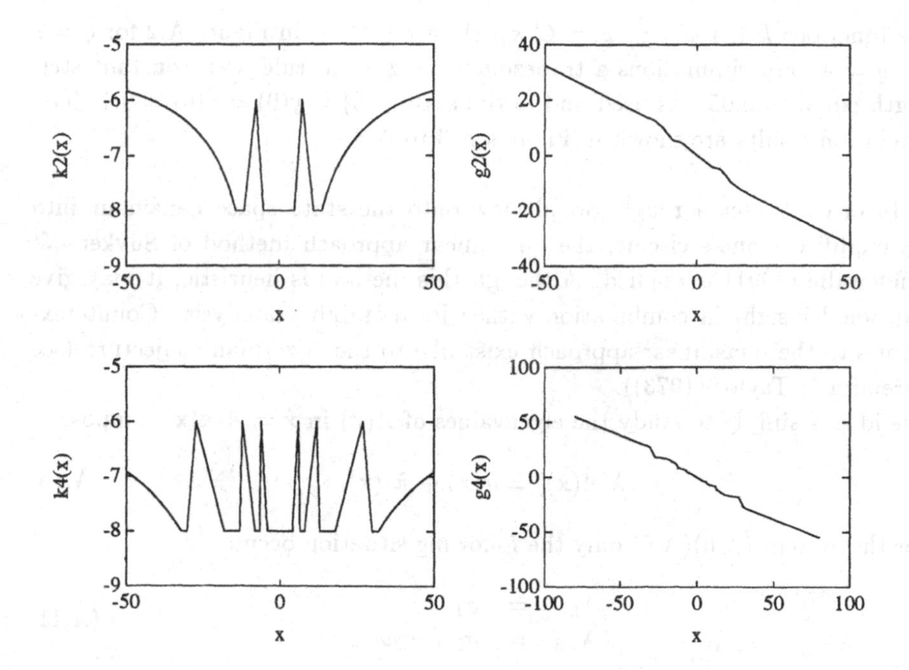


Figure A.2: *Nonlinearities* $k_2(x)$, $k_4(x)$ *and corresponding* $g_2(x) = C_1 k_2(x)x$, $g_4(x) = C_1 k_4(x)x$ *for generating the 2-double and 4-double scroll attractor respectively.*

## A.2 $n$-double scrolls

In order to generate the $n$-double scroll attractors ($n = 1, 2, 3, 4, ...$) the parameter $a$ is fixed at $a = 7$. The attractors are obtained by setting $\gamma_{j,j-1} = 5$ (for $j = 2 + 3(i-2)$, $i = 2, ..., q$). Some examples are listed below

$$\begin{array}{ll}
\text{case } q = 2: & \Delta_2 = [5\ 7.5\ 11.25\ 14]^T \\
& \rightarrow 2\text{-double} \quad \text{scroll} \\
\text{case } q = 3: & \Delta_3 = [5\ 7.5\ 8\ 10\ 15\ 16\ 18]^T \\
& \rightarrow 3\text{-double} \quad \text{scroll} \\
\text{case } q = 4: & \Delta_4 = [4\ 6\ 6.5\ 8\ 12\ 13\ 18\ 27\ 30\ 32]^T \\
& \rightarrow 4\text{-double} \quad \text{scroll} \\
\text{case } q = 6: & \Delta_6 = [4\ 6\ 6.4\ 8\ 12\ 12.8\ 16\ 24\ 25.6\ 32\ 48\ 51.3\ 62\ 93\ 99.4\ 110]^T \\
& \rightarrow 6\text{-double} \quad \text{scroll.}
\end{array} \tag{A.9}$$

The functions $k_q(x)$ and $g_q(x) = C_1 k_q(x) x$ are plotted in Figure A.2 for $q = 2$ and $q = 4$. For simulations a trapezoidal integration rule with constant step length equal to 0.05 was used and initial state $x(0) = y(0) = z(0) = 0.1$. The simulation results are shown in Figures A.3 to A.8.

In order to get a rough global view onto the state space behaviour into the modified Chua's circuit, the quasilinear approach method of Suykens & Vandewalle (1991) is applied. Although this method is heuristic, it may give additional insight in combination with a local stability analysis. Counterexamples to the quasilinear approach exist like to the Aizerman conjecture (see Narendra & Taylor (1973)).
The idea is simply to study the eigenvalues of $A(\mathbf{x})$ in $\dot{\mathbf{x}} = A(\mathbf{x})\mathbf{x}$. Suppose

$$\lambda(A(\mathbf{x})) = \sigma(\mathbf{x}) + j\omega(\mathbf{x}). \tag{A.10}$$

For the system (A.5)(A.6) only the following situation occurs

$$\begin{array}{rcl} \lambda_1 & = & \sigma_1 \\ \lambda_{2,3} & = & \sigma_{2,3} \pm j\omega. \end{array} \tag{A.11}$$

Regions $S, U_r, U_i \subset \mathbb{R}^3$ are now defined as

$$\begin{array}{lcl} \text{if } \sigma_1(\mathbf{x}_P) < 0 \text{ and } \sigma_{2,3}(\mathbf{x}_P) < 0 & \rightarrow & x_P \in S \\ \text{if } \sigma_1(\mathbf{x}_P) > 0 \text{ and } \sigma_{2,3}(\mathbf{x}_P) < 0 & \rightarrow & x_P \in U_r \\ \text{if } \sigma_1(\mathbf{x}_P) < 0 \text{ and } \sigma_{2,3}(\mathbf{x}_P) > 0 & \rightarrow & x_P \in U_i. \end{array} \tag{A.12}$$

In Figure A.10 these regions are plotted for the cases $q = 2$ and $q = 4$ as a function of the bifurcation parameter $a$ between 3 and 8. Together with the local stability analysis, it becomes clear then that one bifurcates from $n$-double sinks to $n$-double scrolls (Figure A.9), like this is the case for the original Chua circuit.

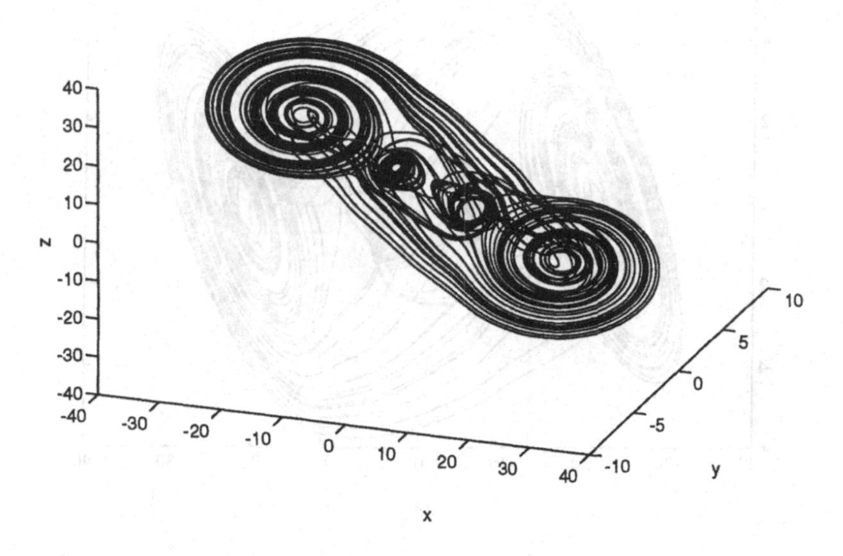


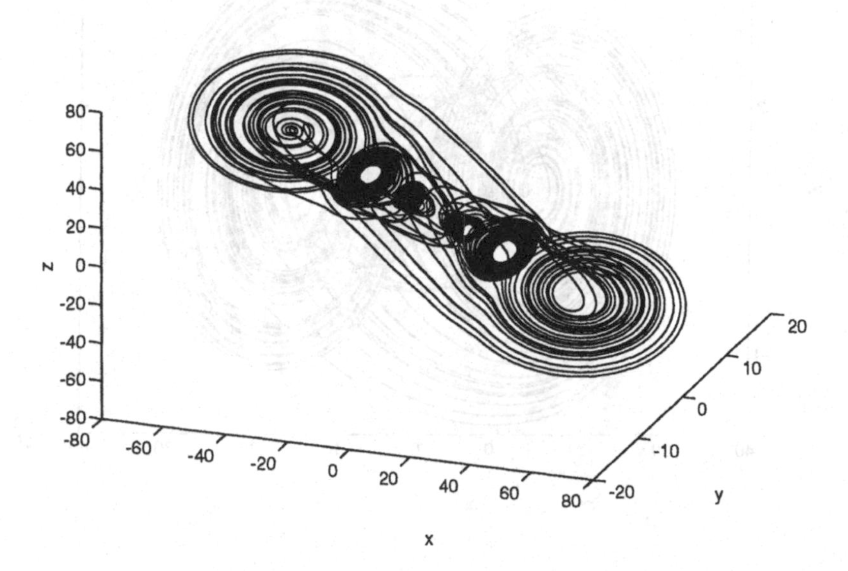


Figure A.3: *A three dimensional view on the 2-double scroll and the 4-double scroll.*

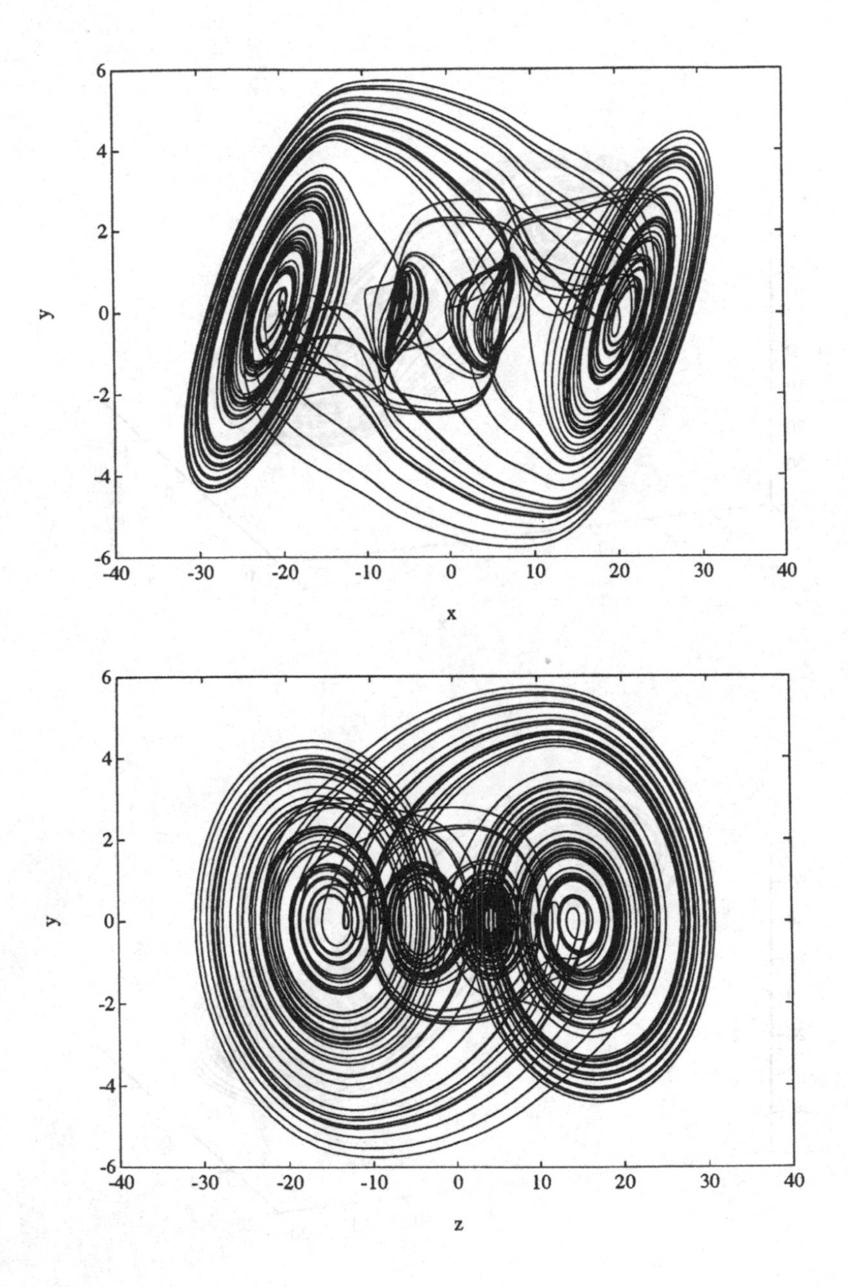


Figure A.4: 2-*double scroll attractor: (Top)* $(x, y)$, *(Bottom)* $(z, y)$.

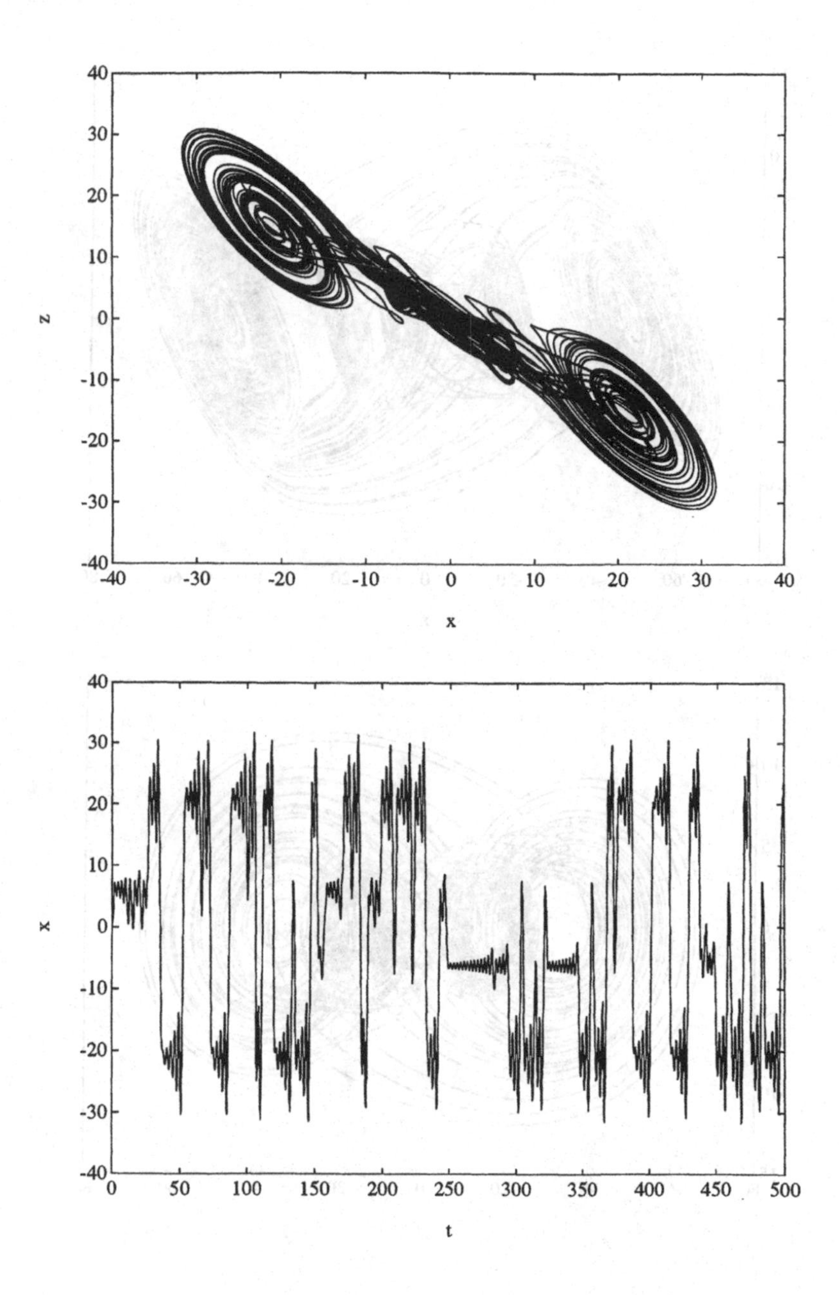


Figure A.5: 2-*double scroll attractor: (Top)* $(x, z)$, *(Bottom)* $x(t)$.

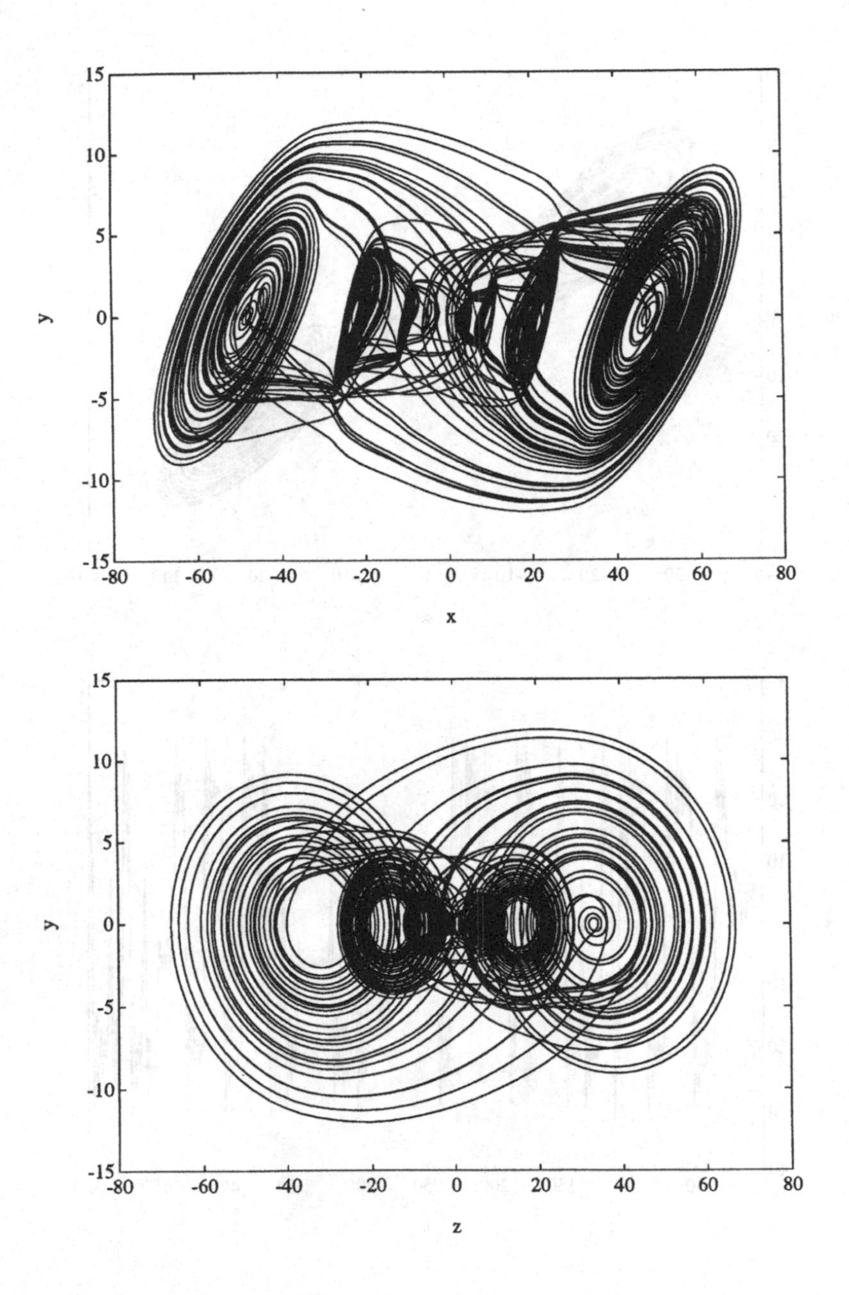


Figure A.6: 4-*double scroll attractor: (Top)* $(x, y)$, *(Bottom)* $(z, y)$.

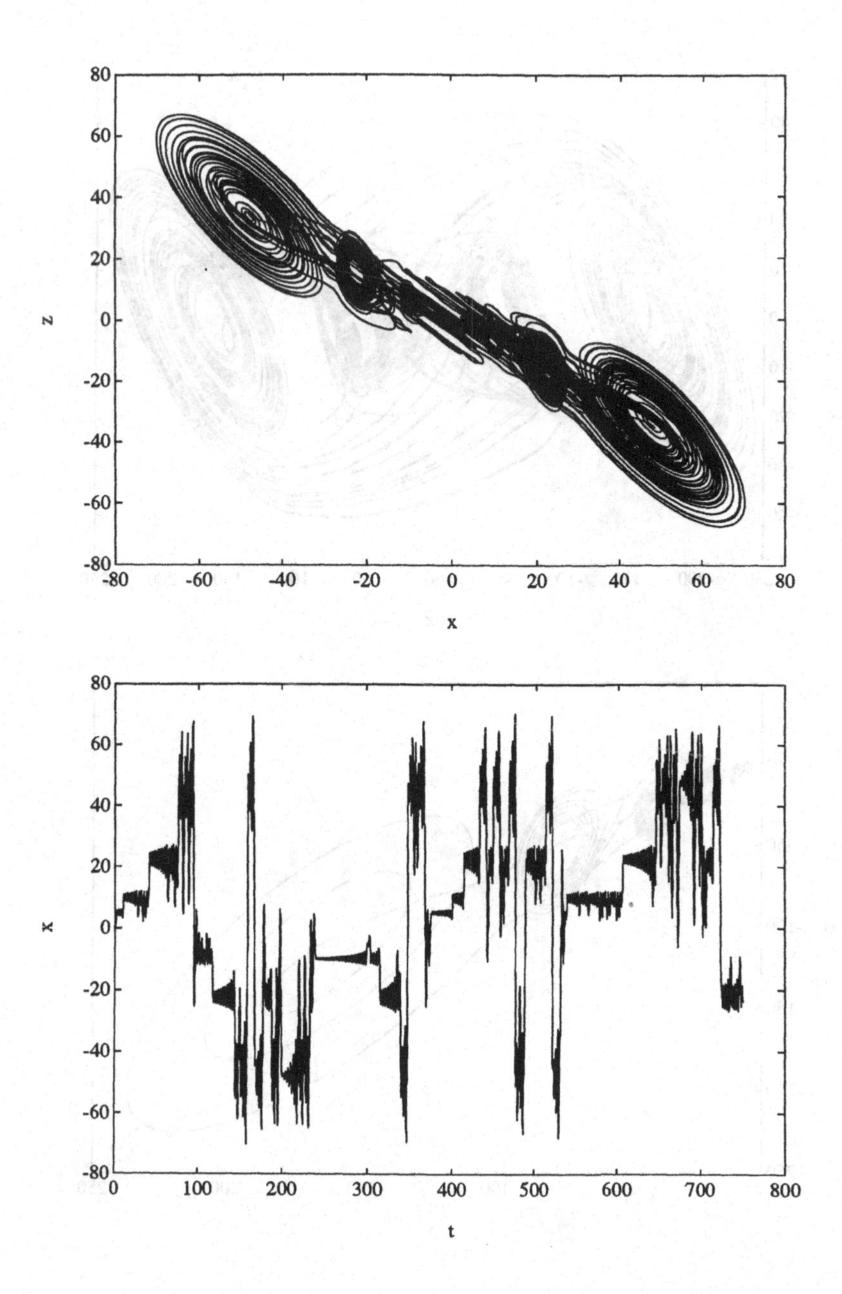


Figure A.7: 4-*double scroll attractor: (Top)* $(x, z)$, *(Bottom)* $x(t)$.

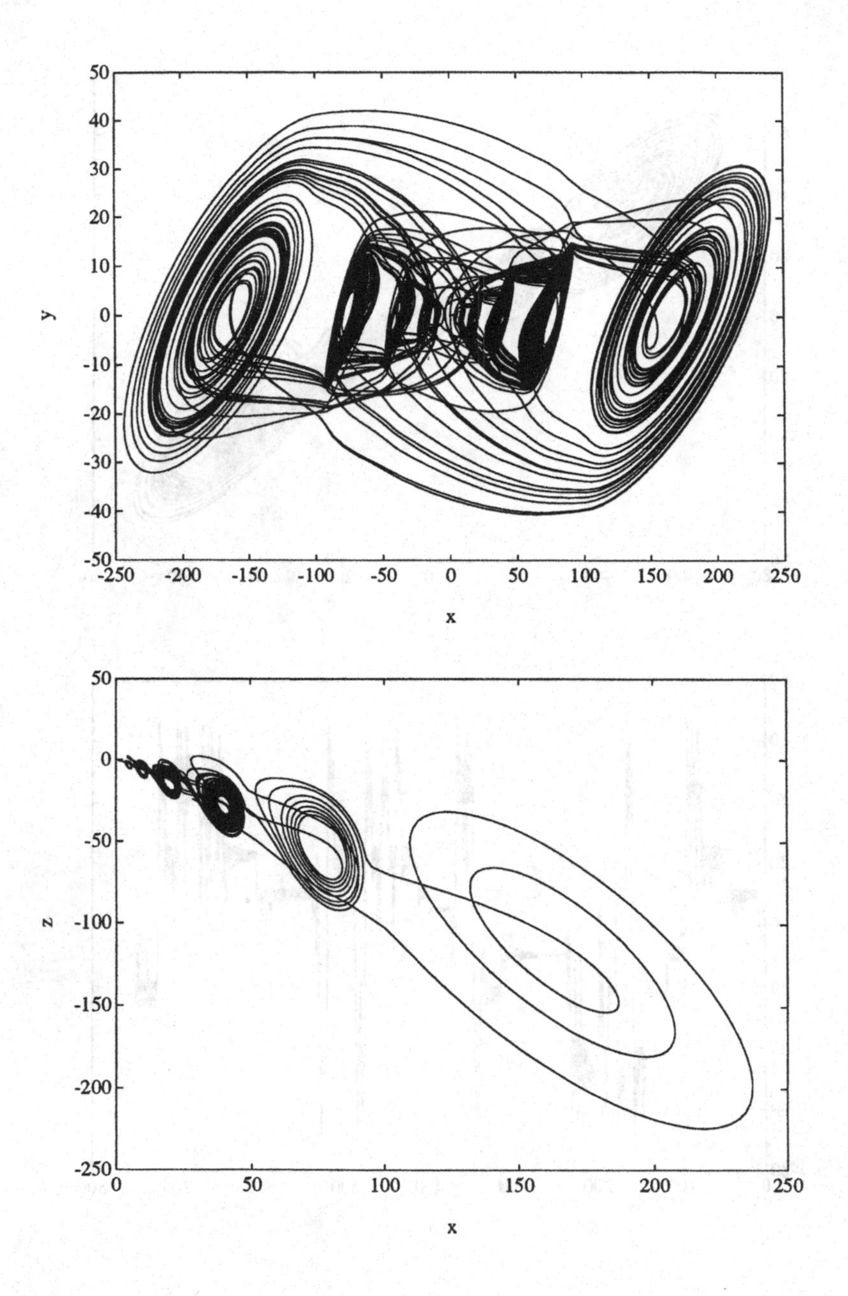


Figure A.8: *6-double scroll attractor: (Top) $(x, y)$; (Bottom) enlarged part of the $(x, z)$ view, showing the 6 scrolls. Note that this Figure is symmetric with respect to the origin.*

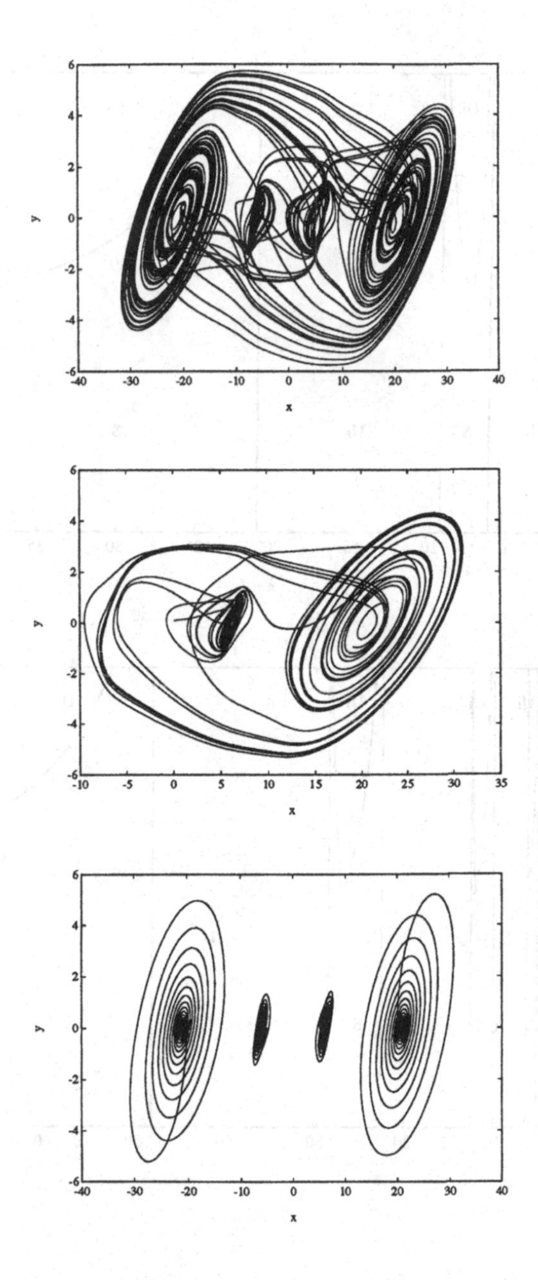


Figure A.9: *Bifurcation from a 2-double sink to a 2-double scroll. (Top) 2-double scroll for* $a = 7$*; (Middle)* $a = 6.5$*; (Bottom) 2-double sink for* $a = 4$.

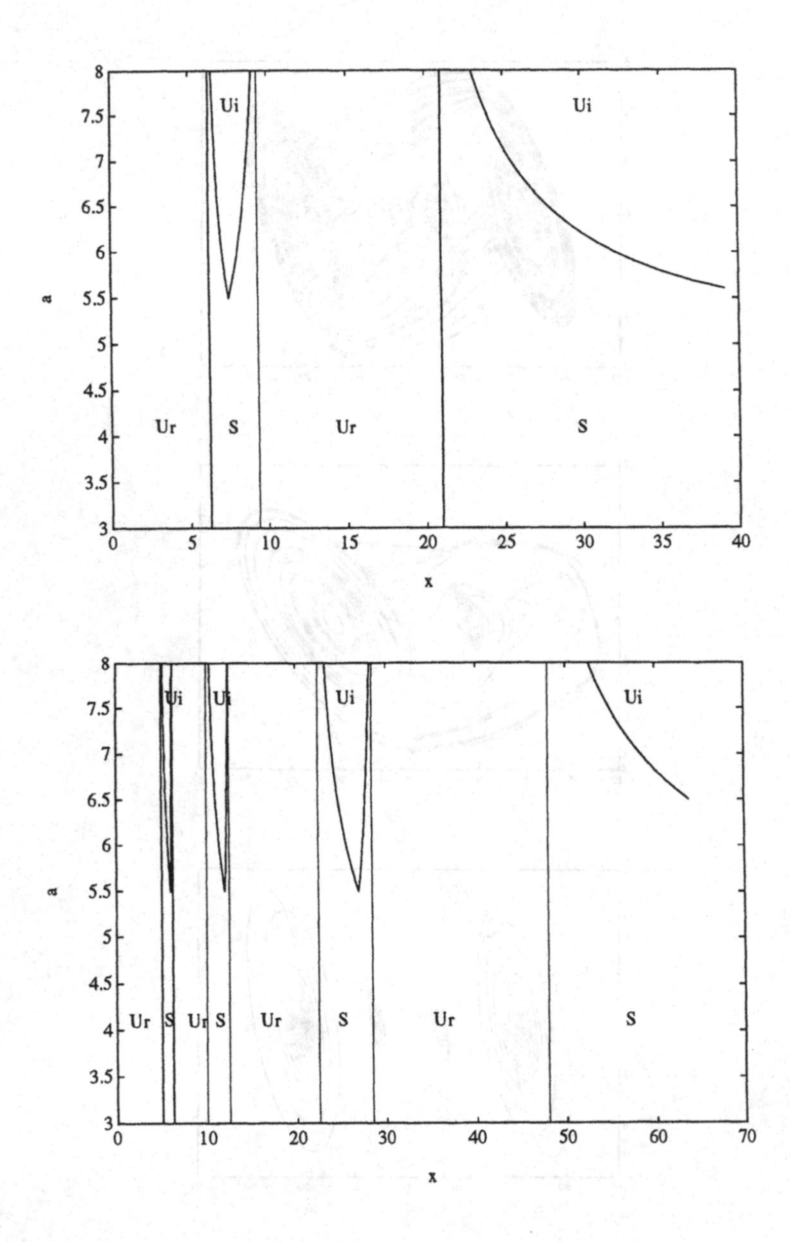


Figure A.10: *Between n-double sinks and n-double scrolls. (Top) $q = 2$; (Bottom) $q = 4$. Slices at $a = 4$ and $a = 7$ correspond respectively to n-double sinks and to n-double scrolls.*

# Appendix B

# Fokker-Planck Learning Machine for Global Optimization

Most problems that we have investigated in this book lead to nonconvex optimization problems and in general many problems arising in learning, classification, identification, control etc. are somehow related to the minimization of a cost function with several local optima. A new method for global optimization is proposed here that is closely related to continuous simulated annealing (or recursive stochastic algorithms in discrete time), according to Suykens & Vandewalle (1994i,j). To these recursive algorithms stochastic differential equations of the Langevin-type are associated. The Fokker-Planck equation for the conditional transition density related to the stochastic differential equation will be studied in order to solve general global differentiable optimization problems. An approximate solution to this Fokker-Planck equation is sought by parametrizing the density with Gaussian sum approximations (radial basis function networks). Like in genetic algorithms a population of points is considered. At each generation the points are generated based upon the transition density and the density is updated according to the Fokker-Planck equation. This update involves basically the solution to a constrained linear least squares problem. The evolution of the Fokker-Planck equation is driven by the ensemble of local geometries at the sampling points, characterized by the gradient of the cost function and the diagonal elements of the Hessian.

This Appendix is organized as follows. In Section B.1 the Fokker-Planck equation related to recursive stochastic algorithms and continuous simulated annealing is derived. In Section B.2 the transition density that satisfies the FP equation is parametrized by means of RBF networks. This leads to the Fokker-Planck machine for global optimization in Section B.3. The Fokker-Planck machine is applied to the neural optimal control problem of swinging up an inverted pendulum in Section B.4.

## B.1 Fokker-Planck equation for recursive stochastic algorithms

Annealing methods were originally applied to finite optimization problems (Kirkpatrick *et al.* (1983)), based on simulating a finite-state Metropolis-type Markov chain. More recently, methods for continuous optimization were developed: in the work of Gelfand & Mitter (1991)(1993) recursive stochastic algorithms were studied for global optimization of a smooth real-valued function $U(.)$ on $\mathbb{R}^n$

$$\min_{x \in \mathbb{R}^n} U(x). \tag{B.1}$$

The algorithms are of the form

$$x_{k+1} = x_k - a_k(\nabla U(x_k) + \xi_k) + b_k w_k \tag{B.2}$$

where $\{w_k\}$ is a sequence of independent $n$-dimensional Gaussian random variables, $\{\xi_k\}$ is a sequence of $\mathbb{R}^n$ valued random variables and $a_k = A/k$, $b_k = \sqrt{B}/\sqrt{k \log\log k}$ for $k$ large and with $A$ and $B$ positive constants. An algorithm like (B.2) was first proposed in Kushner (1987). It is shown that (B.2) converges in probability to the global minima of $U(.)$ under certain additional conditions on $U(.)$, $\{w_k\}$ and $\{\xi_k\}$. The analysis is based on the asymptotic behaviour of the following SDE associated to (B.2), which is a Langevin-type Markov diffusion

$$dx(t) = -\nabla U(x(t))\, dt + \sigma(t)\, dw(t) \tag{B.3}$$

with the scalar $\sigma^2(t) = \sigma_0/\log(t)$ for $t$ large. (B.3) has been called also continuous simulated annealing, the temperature in the cooling scheme is $T(t) = \sigma^2(t)/2$ and $U(x)$ the energy of state $x$. Furthermore the link between (B.2) and (B.3) is established by the relation $\sigma_0 = B/A$. For a fixed temperature $T$ the diffusion has a Boltzmann distribution which is proportional to $\exp(-U(x)/T)$, like for a Metropolis chain.

Now it is well known from the theory of stochastic differential equations that given a Markov process described by

$$dx = f(x,t)\, dt + \sigma(x,t)\, dw \tag{B.4}$$

with $w(.)$ an $\mathbb{R}^m$ valued Wiener process and $\sigma(.) \in \mathbb{R}^{n\times m}$, the conditional transition density $p(x,t|x_0,t_0)$ satisfies the partial differential equation

$$\frac{\partial p}{\partial t} = -\sum_i \frac{\partial}{\partial x_i}(pf_i) + \frac{1}{2}\sum_i\sum_j \frac{\partial^2}{\partial x_i \partial x_j}(p\beta_{ij}) \tag{B.5}$$

with $\beta(x,t) = \sigma(x,t)\sigma(x,t)^T$. Here $p(x,t|x_0,t_0)$ denotes the probability density of being in state $x$ at time $t$ given the process is in state $x_0$ at $t_0$ and the PDE is called the *Fokker-Planck equation* or *forward Kolmogorov equation.* In (B.4) $f$ is called drift and $\sigma$ fluctuation. The solution corresponding the initial condition $p(x,t|x_0,t_0) = \delta(x - x_0)$ is called the fundamental solution. This initial condition means that one is certain that the initial state is equal to $x_0$. Existence and uniqueness of solutions to (B.5) are guaranteed under certain conditions of continuity and restriction on growth of $f$ and $\sigma$. A comprehensive introduction to the theory can be found e.g. in Åström (1970), Arnold (1974) or Bryson & Ho (1969). A rigorous treatment is given e.g. in Doob (1953), Gihman & Skorohod (1979), Ikeda & Watanabe (1981) and Wong (1971).

It is then clear that the Langevin-type diffusion (B.3) is a special case of the SDE (B.4) with $f(x,t) = -\nabla U(x)$, $\sigma(x,t)$ a diagonal matrix and $m = n$. The Fokker-Planck equation related to (B.3) becomes then

$$\frac{\partial p}{\partial t} = \sum_i \frac{\partial U}{\partial x_i}\frac{\partial p}{\partial x_i} + \sum_i \frac{\partial^2 U}{\partial x_i^2} p + \frac{1}{2}\sigma^2(t)\sum_i \frac{\partial^2 p}{\partial x_i^2}. \tag{B.6}$$

Important to notice is that the evolution of the transition density is determined by the geometry of the cost function $U$, characterized by the gradient and the diagonal elements of the Hessian instead of by the function values of the cost function in itself. The fact that the off-diagonal elements of the Hessian are not needed is interesting from a computational point of view, because in a higher dimensional search space the computation of the full Hessian becomes infeasible. The latter is indeed the underlying reason for the use of quasi-Newton methods in nonlinear programming, where one tries to build up curvature information of the Hessian based upon first order gradient information only.

Some examples from physics with analytical solutions to (B.6) are (see e.g. Wang & Uhlenbeck and Kac in Wax (1954))

1. *Constant cost function U*: This is the case of a zero drift term. An example is the motion of a free particle in a surrounding medium, a problem studied by Einstein. In the one dimensional case (B.6) is then of the form

$$\frac{\partial p}{\partial t} = D\frac{\partial^2 p}{\partial x^2}$$

with $D$ a physical constant. The fundamental solution is given by

$$p(x,t|x_0,t_0) = \frac{1}{(4\pi Dt)^{1/2}} \exp[-(x-x_0)^2/(4Dt)].$$

2. *Quadratic cost function* $U$: In the one dimensional case (B.6) becomes now

$$\frac{\partial p}{\partial t} = \alpha \frac{\partial}{\partial x}(xp) + D \frac{\partial^2 p}{\partial x^2}$$

and is related to the cost function $U(x) = 0.5\,\alpha x^2$. The fundamental solution is given by

$$p(x,t|x_0,t_0) = \frac{1}{(2\pi\sigma_p^2)^{1/2}} \exp[-(x-\overline{x})^2/2\sigma_p^2]$$

where $\overline{x} = x_0 \exp(-\alpha t)$ and $\sigma_p^2 = (D/\alpha)[1-\exp(-2\alpha t)]$. The higher dimensional case corresponds to the Brownian motion of a system of coupled harmonic oscillators and its fundamental solution is also Gaussian and the related SDEs are *linear*. The case of linear SDEs is also extensively studied in system theory (see e.g. Åström (1970), Bryson & Ho (1969)).

## B.2 Parametrization of the pdf by RBF networks

In order to minimize general type of cost functions in possibly high dimensional search spaces, a parametrization of the density $p$ will be taken by means of radial basis function networks. Such a parametrization makes sense because of the universal approximation ability of RBF networks (see e.g. Park & Sandberg (1991), Poggio & Girosi (1990), Broomhead & Lowe (1988)). Because $p$ is a density the following conditions are required

1. $p$ is continuous
2. $\int_{-\infty}^{\infty} \dots \int_{-\infty}^{\infty} p(x)dx = 1$
3. $p(x) \geq 0$ for all $x$

This leads to the parametrization

$$\hat{p}(x,t|x_0,t_0) = \sum_{i=1}^{n_h} w_i(t)\, N(x - s_i(t), R_i(t)), \quad \sum_{i=1}^{n_h} w_i(t) = 1, \quad w_i \geq 0 \qquad (B.7)$$

with $N(s,R) = k|R|^{-1/2}\exp(-\frac{1}{2}s^T R^{-1} s)$, $k = (2\pi)^{-n/2}$. The constraints $\sum_{i=1}^{n_h} w_i(t) = 1$ and $w_i \geq 0$ ensure that the function is a density. Here are $n_h$ the number of hidden neurons, $w_i$ the weights of the output layer, $s_i \in \mathbb{R}^n$ the center and $R_i \in \mathbb{R}^{n\times n}$ the covariance matrix related to the $i$-th hidden neuron. Such a Gaussian sum approximation was already proposed before in the field of nonlinear filtering: nonlinear Bayesian estimation using Gaussian sum approximations is indeed discussed in Alspach & Sorenson (1972), Sorenson & Alspach (1971) and Anderson & Moore (1979). In the sequel we restrict the matrices $R_i^{-1} = \alpha_i^2 I_n$ to be diagonal for computational reasons. In that case

$$\hat{p}(x;\theta(t)) = k \sum_{i=1}^{n_h} v_i(t)\, \exp[h_i(t)] \tag{B.8}$$

with $v_i(t) = w_i(t)[\alpha_i(t)]^n$, $h_i(t) = -\frac{1}{2}[\alpha_i(t)]^2 \sum_{j=1}^{n}[x_j - c_{ji}(t)]^2$ and $\theta \in \mathbb{R}^q$ denotes the parameter vector containing the elements $w_i$, $\alpha_i$ and $c_{ji}$.

In order to determine the unknown parameter vector $\theta$ during time, we will consider a population of $N$ points that samples the state space at a given time $t$ and update the population at each generation like this is done in genetic algorithms (see e.g. Goldberg (1989)). The update is based here upon the density that satisfies the Fokker-Planck equation. The incoming information of

$$\{\frac{\partial U}{\partial x_i}|_{x=x^{(l)}}, \frac{\partial^2 U}{\partial x_i^2}|_{x=x^{(l)}}\}_{l=1,\ldots,N}$$

at the $N$ points $x^{(l)}$ yields an estimate for $\dot{\theta}$. This is done by substituting (B.8) into the equation (B.6) and evaluating the equation at the $N$ points. One obtains

$$\begin{aligned} \frac{\partial \hat{p}}{\partial t}|_{x=x^{(l)}} = \; & \sum_i \frac{\partial U}{\partial x_i}|_{x=x^{(l)}} \frac{\partial \hat{p}}{\partial x_i}|_{x=x^{(l)}} + \sum_i \frac{\partial^2 U}{\partial x_i^2}|_{x=x^{(l)}}\, \hat{p}\,|_{x=x^{(l)}} \cdots \\ & + \frac{1}{2}\sigma^2(t) \sum_i \frac{\partial^2 \hat{p}}{\partial x_i^2}|_{x=x^{(l)}}, \quad l = 1,\ldots,N \end{aligned} \tag{B.9}$$

with

$$\left\{ \begin{aligned} \frac{\partial \hat{p}}{\partial t} &= k\sum_i(\dot{w}_i\alpha_i^n + n w_i \alpha_i^{n-1}\dot{\alpha}_i - v_i\alpha_i\dot{\alpha}_i \textstyle\sum_j z_{ji}^2 + v_i\alpha_i^2 \sum_j z_{ji}\dot{c}_{ji}) \exp(h_i) \\ \frac{\partial \hat{p}}{\partial x_j} &= -k\textstyle\sum_i v_i \alpha_i^2 z_{ji} \exp(h_i) \\ \frac{\partial^2 \hat{p}}{\partial x_j^2} &= -k\textstyle\sum_i v_i\alpha_i^2(1-\alpha_i^2 z_{ji}^2)\exp(h_i) \end{aligned} \right.$$

and by definition $z_{ji} = x_j - c_{ji}$. Assuming that we take $N > q$, i.e. a population size that is larger than the number of parameters in $\theta$, (B.9) is an overdetermined set of linear equations in the unknown $\dot{\theta} = [\dot{w}_i; \dot{\alpha}_i; \dot{c}_{ji}]$. In addition we have to take into account the constraint $\sum_i w_i = 1$ as $\sum_i \dot{w}_i = 0$. This leads finally to the problem

$$A\,\dot{\theta} = b \quad \text{such that} \quad \sum_i \dot{w}_i = 0 \tag{B.10}$$

where the meaning of $A \in \mathbb{R}^{N\times q}$ and $b \in \mathbb{R}^N$ follows immediately from (B.9). In fact the PDE (B.6) is transformed into a nonlinear ODE. Given

$$A(\theta)\,\dot{\theta} = b(\theta), \qquad \theta(0) = \theta_0 \quad \text{such that} \quad c^T\dot{\theta} = 0$$

where $c$ is some constant vector, it can be shown that a solution in least squares sense is given by

$$\dot{\theta} = S(\theta)^{-1}[A(\theta)^T b(\theta) - \frac{c^T S(\theta)^{-1} A(\theta)^T b(\theta)}{c^T S(\theta)^{-1} c}\, c], \qquad \theta(0) = \theta_0 \tag{B.11}$$

with $S(\theta) = A(\theta)^T A(\theta) + \delta I$ and $\delta$ a regularization parameter. (B.11) forms essentially the learning rule for the Fokker-Planck learning machine. Given the current parameter vector of the pdf it describes how the pdf has to change as new information from samples in search space is coming in.

## B.3 FP machine: conceptual algorithm

The previous ideas lead to the following conceptual algorithm for solving a global optimization problem.

- **Parameters:**

| | |
|---|---|
| $n$ : | dimension of search space |
| $N$ : | number of points in the population |
| $n_h$ : | number of hidden neurons in Gaussian sum approximation |
| $q$ : | number of elements of $\theta$. Note that $q = (n+2)n_h$. |
| $n_g$ : | maximum number of generations |
| $\theta(0)$ : | initial parameter vector for $\hat{p}(x;\theta(t))$ |

- **Generation 1** ($t = 0, k = 1$):
  - Choose $\theta(0)$.
  - Generate an initial population of $N$ points ($N > q$) from $\theta(0)$.
  - Choose $\sigma_0$, related to the initial temperature.

- **While** $k \leq n_g$
  1. **Generation** $k$**:**
     - Propose the Gaussian sum approximation $\hat{p}(x;\theta(t))$ (B.8) as solution to the FP equation (B.6).
     - Determine $\dot{\theta}$ based upon the derivatives at the $N$ points of the current population

       $$\{\frac{\partial U}{\partial x_i}|_{x=x^{(l)}}, \frac{\partial^2 U}{\partial x_i^2}|_{x=x^{(l)}}\}_{l=1,\ldots,N}$$

       by solving the constrained linear least squares problem (B.10).
     - Evaluate the cost function at the $N$ points.
  2. **Next generation** ($k := k+1$)
     - Update $\theta$:

       $$\theta(t+dt) = \theta(t) + \dot{\theta}\, dt,$$

       by choosing an appropriate step size $dt$. Set $t := t + dt$.
     - In case there exists a $w_i \leq 0$, then remove the $i$-th hidden neuron ($n_h := n_h - 1$).
     - Update the population, based upon the marginal densities of $\hat{p}(x;\theta(t))$ (it is explained in (B.12) how these densities are generated).
     - Update the temperature, according to the cooling scheme in (B.3).

**End**

**Remarks.**

- The initial density does not correspond here to a Dirac impulse. Hence we are not looking for the fundamental solution. For a constant temperature one obtains a Boltzmann distribution for $\lim_{t\to\infty} p(x,t)$, which is proportional to $\exp[-U(x)/T]$. The algorithm is mainly intended for localization of 'good' local optima. After detecting a 'good region' by means of the FP-algorithm, one can switch to a local optimization method (such as a quasi-Newton or a conjugate gradient method) for fast convergence to the optimum. In the conceptual algorithm, (B.11) is solved using a forward Euler method. More sophisticated integration rules can be used here.

- One of the advantages of the Gaussian sum approximation is that the marginal densities are immediately obtained from the transition density $\hat{p}(x;\theta)$ as

$$\hat{p}(x_j;\theta) = k \sum_{i=1}^{n_h} v_i \exp[-\frac{1}{2}\alpha_i^2(x_j - c_{ji})^2], \quad j = 1,..,n. \tag{B.12}$$

The $N$ points of the population are then generated from these marginal densities. Therefore the following procedure is taken. First a random number is generated in the interval $[0,1]$ according to a uniform distribution. Because $\sum_{i=1}^{n_h} w_i = 1$, one looks to which of the intervals $[\sum_{i=1}^{r} w_i, \sum_{i=1}^{r+1} w_i]$ (with $r = 1, ..., n_h - 1$) this random number belongs. Suppose this is for $r = s$, then the $x_j$ is equal to a random variable, generated according to a Gaussian distribution with mean $c_{js}$ and variance $1/\alpha_s$. This is repeated for all the components $x_j$ of the $N$ points.
Another advantage of the Gaussian sum is that centers are easily removed or added, while preserving the property that the function is a density.

- One has to take into account some constraints with respect to the choice of $n_h$ and $N$. A first constraint is related to the constrained linear least squares problem: in order to have an overdetermined set of equations:

$$(n+2)\, n_h < N.$$

A second one is due the limited number of function evaluations $n_{f_{max}}$ that is allowed in practice

$$5\, n\, n_g\, N < n_{f_{max}}.$$

It is assumed here that an evaluation of the gradient and the second order derivative involves twice $2nN$ function evaluations and that the cost function is also evaluated at the N points at each generation. This yields approximately $5nN$ number of function evaluations per generation. From this expression it is clear that there exists a trade-off between $n_g$ and $N$: either one can run small populations over a large number of generations or large populations over a smaller number of generations. However it follows from multidimensional sampling theory (see Sanner & Slotine (1992) and Slotine & Sanner (1993)) that there exists some lower bound on the number of points $N$ to be taken at a certain generation $k$. This minimal number of points is related to the 'bandwidth' of $U(x)$. The larger this bandwidth the more points should be taken.
A qualitative explanation can be given as follows. Let us first assume

that there is persistence of excitation. Roughly speaking this means that the search space is sampled sufficiently well in all regions for which we want to have a good approximation, in order to obtain a small modeling error between $\hat{p}(x)$ and $p(x)$, together with a good generalization (see Slotine & Sanner (1993) for a more precise statement). In order to have then an exact reconstruction of $p(x)$ based upon samples of $p(x)$ only, the original function $\hat{p}(x)$ should be sampled at a spatial sampling frequency which is 'high enough' (suppose the modeling error between $p$ and $\hat{p}$ can be made arbitrarily small, such that we can replace $p(x)$ by $\hat{p}(x)$). In the case of a rectangular lattice (instead of the randomly generated points as considered in the conceptual algorithm) this statement can be made more precise (see Sanner & Slotine (1992)): consider the lattice

$$\xi_I = i_1 \, \Delta \, e_1 + ... + i_n \, \Delta \, e_n$$

where $I = \{i_1, ..., i_n\}$ is an $n$-tuple of integers and $\Delta$ is the sample spacing along the lattice basis vectors $\{e_i\}_{i=1}^n$ (suppose a standard basis for $\mathbb{R}^n$). Now suppose we would know the exact value of $p(x)$ at those points, in other words suppose we know $p_s(x) = \sum_{I \in \mathbb{Z}^n} \delta(x - \xi_I) p(x)$. Furthermore suppose that the spatial Fourier transform $P(\nu)$ of $p(x)$

$$P(\nu) = \int_{\mathbb{R}^n} p(x) \exp(-2\pi j \nu^T x) dx$$

has compact support and let $K(\beta)$ be the smallest $n$-cube $[-\beta, \beta]^n$, centered at the origin that encloses the support of $P(\nu)$. In order to reconstruct then $p(x)$ perfectly from $p_s(x)$ the following condition should hold: $\Delta \leq 1/(2\beta)$ (For $n = 1$ this corresponds to the well known Nyquist sampling Theorem). The link between $p_s(x)$ and the sampling of $U(x)$ can be understood in the case of a constant temperature for which $\lim_{t \to \infty} p(x, t)$ is proportional to $\exp[-U(x)/T]$.

## B.4 Examples

In a first example we consider the minimization of the cost function

$$U(x) = \frac{1}{2n} \sum_{i=1}^{n} x_i^2 - 4n \prod_{i=1}^{n} \cos(x_i) + 4n$$

with $n = 10$, which has many local optima. This example was already described in Styblinski & Tang (1990). In Suykens & Vandewalle (1994i) a comparison is made between the following algorithms:

- FSA: Fast simulated annealing (Szu (1987)), see Styblinski & Tang (1990).
- SAS: Stochastic approximation with function smoothing (Styblinski & Tang), see Styblinski & Tang (1990).
- Multistart QN: Multistart local optimization scheme with a quasi-Newton method.
- FP: Fokker-Planck Learning Machine (Suykens & Vandewalle (1994)).
- FP + QN: Fokker-Planck Learning Machine + additional local optimization with quasi-Newton method starting from the best point, delivered by FP (Suykens & Vandewalle (1994)).

Simulation results indicate (after 10 runs of the algorithms) that best results were obtained by means of FP+QN and SAS, in terms of the quality of the obtained solution and the number of function evaluations. The SAS algorithm has however the disadvantage of many tuning parameters. Futhermore FP+QN was considerably better than multistart QN. FP and FSA had comparable results. Hence one may conclude that a good procedure is to apply first FP in order to locate 'good regions' and finally to apply QN for fast convergence to the local optimum.

In a second example we consider again the swinging up problem of an inverted pendulum system, controlled by a multilayer feedforward neural controller with 4 hidden neurons. The cost function is the norm of the state vector at the endpoint of the simulation of the closed loop system. The same parameters are taken as in Chapter 4.

For the FP algorithm a parameter vector $par = [\sigma_0; n_h; N; n_g; k_1; k_2; dt]$ is defined. Here $k_1$ is related to the initial standard deviations in the Gaussian sum approximation ($k_1 = 1/\alpha_i(0)(i = 1, ..., n_h)$). The centers $c_{ji}$ of the initial Gaussian sum are random uniformly distributed in the interval $[-k_2, k_2]$. The '*par*'-vector is taken here equal to $par = [1; 3; 100; 15; 0.1; 0.1; 1]$. The first and second order derivatives of the cost function were calculated numerically (forward difference). The simulation results are shown in Table B.1. The method FP+SQP in the table means the local optimization scheme SQP (Sequential Quadratic Programming for constrained optimization) applied to the best point delivered by the FP-algorithm. SQP was used with random initial points uniformly distributed in $[-k_3, k_3]^n$ with $k_3 = 0.2$. Ten runs were done for the FP-algorithm. For SQP one run was done which consists of 60 starting points. The lowest cost function that was obtained by FP is equal to $9.5217\,e-08$, which comes very close to the theoretical lower bound which is 0 because the cost function is a norm. From the simulation results is clear that FP+QN is considerably better than multistart QN.

| | $U_{max}$ | $U_{min}$ | $E\{U\}$ | $\sigma(U)$ | $E\{n_f\}$ |
|---|---|---|---|---|---|
| FP | $1.0923e+00$ | $9.0800e-02$ | $4.8130e-01$ | $4.1320e-01$ | 120000 |
| FP + SQP | $7.5763e-02$ | $9.5217e-08$ | $1.2822e-02$ | $2.4415e-02$ | 122000 |
| SQP | $1.3600e+03$ | $3.7131e-05$ | $5.7233e+02$ | $6.7163e+02$ | 120000 |

Table B.1: *Comparison of FP, FP+SQP and multi start SQP on the neural optimal control problem of swinging up an inverted pendulum system by means of a feedforward neural controller (16 unknowns) of Chapter 4. The cost function is the norm of the state vector at the endpoint of the simulation of the closed loop system. $U_{max}$, $U_{min}$, $E\{U\}$, $\sigma(U)$ and $E\{n_f\}$ mean respectively the maximum, minimum, mean and standard deviation of the cost function and the mean of the number of function evaluations over the 10 runs (and the 1 run in the SQP case).*

# B.5 Conclusions

A new algorithm for solving global optimization problems has been proposed which follows from considering a Fokker-Planck equation that is related to recursive stochastic algorithms. The Fokker-Planck PDE was transformed into a nonlinear ODE by parametrizing the transition density with Gaussian sum approximations. The algorithm works with a population of points (like GAs). At each generation a population is generated based upon the marginal densities related to the Gaussian sum approximation and the population is updated according to the density that satisfies the Fokker-Planck equation. It is illustrated with examples that the algorithm is particularly interesting for locating regions of good local optima and is especially successful in combination with a fast local optimization scheme.

We have studied parametrizations by means of Gaussian sum approximations (RBFs) for the FP equation. Due to these approximations it was easy to update statistics, to generate marginal distributions, to remove or add centers and to impose the condition that the function has to be a density, advantages which were already exploited in the field of nonlinear filtering (Alspach & Sorenson (1972), Anderson & Moore (1979)). However as suggested also in Sanner & Slotine (1992) it might be worthwhile to study other parametrizations such as Gabor functions or wavelets which have a better combined space-frequency localization.

The Fokker-Planck equation for global optimization is driven by the gradient and the diagonal elements of the Hessian of the cost function, and hence driven by the geometry of the cost function instead of by function values (or fitness values like this is called in GA). With respect to aspects of geometry, application of results from information geometry (Amari (1992)) and stochastic differential geometry (see e.g. Ikeda & Watanabe (1981), Rogers & Williams (1987)) may possibly lead to better insight of the working principles and understanding of limitations of the method. In Amari (1992) this led to new insight and learning rules for Boltzmann machines. Other related work is e.g. the Boltzmann g-RHONN, a learning machine for estimating unknown probability distributions (Kosmatopoulos & Christodoulou (1994)). A gradient recurrent high-order neural network is proposed there in order to approximate the stationary density for the SDE. The FP equation was however only indirectly mentioned.

# Appendix C

# Proof of $\mathbf{NL}_q$ Theorems

## Proof of Lemma 5.1

This is shown for each model $\mathcal{M}_i$ in particular

1. Model $\mathcal{M}_0$: $q = 1$ by taking $\Gamma_1 = I$, $V_1 = A$, $B_1 = [B\ K\ 0]$, $\Lambda_1 = I$, $W_1 = C$, $D_1 = [D\ I\ 0]$, $p_k = x_k$, $w_k = [u_k; \epsilon_k; 1]$.

2. Model $\mathcal{M}_1$: $q = 2$ by taking $\Gamma_1 = I$, $\Gamma_2 = \Gamma_{AB}$, $V_1 = W_{AB}$, $V_2 = V_A$, $B_2 = [V_B\ 0\ \beta_{AB}]$, $B_1 = [0\ K\ 0]$, $\Lambda_1 = I$, $\Gamma_2 = I$, $W_1 = I$, $W_2 = C$, $D_2 = [D\ I\ 0]$, $D_1 = I$. $\Gamma_{AB}$ is a diagonal matrix with diagonal elements $\in [0, 1]$, like this is shown for (5.8)-(5.9).

3. Model $\mathcal{M}_2$: $q = 2$ by taking $\Gamma_1 = I$, $\Gamma_2 = \Gamma_{AB}$, $V_1 = W_{AB}$, $V_2 = V_A$, $B_2 = [V_B\ 0\ \beta_{AB}]$, $B_1 = [0\ K\ 0]$, $\Lambda_1 = I$, $\Lambda_2 = \Gamma_{CD}$, $W_1 = W_{CD}$, $W_2 = V_C$, $D_2 = [V_D\ 0\ \beta_{CD}]$, $D_1 = [0\ I\ 0]$.

4. Model $\mathcal{M}_3$: $q = 3$ by taking $\Gamma_1 = I$, $\Gamma_2 = \Gamma_{AB_w}$, $\Gamma_3 = \Gamma_{AB_v}$, $V_1 = W_{AB}$, $V_2 = V_{AB}$, $V_3 = V_A$, $B_3 = [V_B\ 0\ \beta_{AB_v}]$, $B_2 = [0\ 0\ \beta_{AB_w}]$, $B_1 = [0\ K\ 0]$, $\Lambda_1 = I$, $\Lambda_2 = \Gamma_{CD_w}$, $\Lambda_3 = \Gamma_{CD_v}$, $W_1 = W_{CD}$, $W_2 = V_{CD}$, $W_3 = V_C$, $D_3 = [V_D\ 0\ \beta_{CD_v}]$, $D_2 = [0\ 0\ \beta_{CD_w}]$, $D_1 = [0\ I\ 0]$.

## Proof of Theorem 5.1

Propose a Lyapunov function $V(p) = \|D_1 p\|_2$ (positive, radially unbounded and $V(0) = 0$) with $D_1 \in \mathbb{R}^{n_p \times n_p}$ a diagonal matrix with nonzero diagonal elements. Hence $V_k = \|D_1 p_k\|_2$ and $V_{k+1} = \|D_1 \Gamma_1 V_1 \Gamma_2 V_2 ... \Gamma_q V_q p_k\|_2$. The following procedure is applied then: (a) insert $D_i^{-1} D_i$ after $\Gamma_i$ where

$D_i \in \mathbb{R}^{n_{h_i} \times n_{h_i}} (i = 1, ..., q-1)$ and insert $D_1^{-1} D_1$ after $V_q$; (b) the diagonal matrices $\Gamma_i$ and $D_i^{-1}$ commute; (c) use the properties of induced norms and the fact that $||\Gamma_i||_2 \leq 1$. Hence

$$\begin{aligned} V_{k+1} &= ||D_1 \Gamma_1 D_1^{-1} D_1 V_1 ... \Gamma_q D_q^{-1} D_q V_q D_1^{-1} D_1 p_k||_2 \\ &\leq ||\Gamma_1||_2 ||D_1 V_1 D_2^{-1}||_2 ||\Gamma_2||_2 ||D_2 V_2 D_3^{-1}||_2 ... ||\Gamma_q||_2 ||D_q V_q D_1^{-1}||_2 V_k \\ &\leq ||D_1 V_1 D_2^{-1}||_2 ||D_2 V_2 D_3^{-1}||_2 ... ||D_q V_q D_1^{-1}||_2 V_k. \end{aligned}$$

Defining $\beta_D = \prod_{i=1(\text{moduloq})}^{q} ||D_i V_i D_{i+1}^{-1}||_2$, a sufficient condition for global asymptotic stability of the $NL_q$ system is $\beta_D < 1$, because then $\Delta V_k = V_{k+1} - V_k < (\beta_D - 1) V_k < 0$.
This condition $\beta_D < 1$ is satisfied if

$$\max_i \{||D_i V_i D_{i+1}^{-1}||_2 : i = 1, ..., q \ (\text{mod} q)\} \leq \beta_D^{1/q} < 1$$

or if

$$\overline{\sigma}(\begin{bmatrix} D_1 V_1 D_2^{-1} & & & \\ & D_2 V_2 D_3^{-1} & & \\ & & \ddots & \\ & & & D_q V_q D_1^{-1} \end{bmatrix}) \leq \beta_D^{1/q} < 1$$

or if

$$\overline{\sigma}(\begin{bmatrix} & I_{n_{h_2}} & & \\ & & \ddots & \\ & & & I_{n_{h_q}} \\ I_{n_{h_1}} & & & \end{bmatrix} \cdot \begin{bmatrix} D_1 V_1 D_2^{-1} & & & \\ & D_2 V_2 D_3^{-1} & & \\ & & \ddots & \\ & & & D_q V_q D_1^{-1} \end{bmatrix}) \leq \beta_D^{1/q} < 1.$$

The condition on this permuted block diagonal matrix is then satisfied if

$$||D_{tot} V_{tot} D_{tot}^{-1}||_2^q \leq \beta_D < 1$$

where $D_{tot} = \text{diag}\{D_2, D_3, ..., D_q, D_1\}$.

## Proof of Theorem 5.2

Propose a Lyapunov function $V(p) = ||P_1 p||_2$ (positive, radially unbounded and $V(0) = 0$) with $P_1 \in \mathbb{R}^{n_p \times n_p}$. Hence $V_k = ||P_1 p_k||_2$ and $V_{k+1} = ||P_1 \Gamma_1 V_1 \Gamma_2 V_2 ... \Gamma_q V_q p_k||_2$. Defining $q_k = V_1 \Gamma_2 V_2 ... \Gamma_q V_q p_k$, we first study under what conditions there exist a matrix $P_1$ and a positive scalar $\alpha_1$ such that the following is satisfied: $\forall q$ and $\forall \Gamma_1$ ($\gamma_{1_i} \in [0, 1]$):

$$||P_1 \Gamma_1 q_k||_2^2 \leq (1 + \alpha_1) ||P_1 q_k||_2^2$$

or $(1+\alpha)S - \Gamma S \Gamma > 0$, where $S = P_1^T P_1$, $\Gamma = \Gamma_1$ (index 1 is omitted for notational reasons). By applying a congruence transformation, the condition

$$(1+\alpha)S - \Gamma S \Gamma > 0$$

is equivalent to

$$(1+\alpha)U - \Gamma U \Gamma > 0,$$

because $\Gamma$, $N^{-1}$ and $N$ commute if one defines $U = N^{-1}SN$, with $N$ a diagonal matrix with positive diagonal elements.
Defining then $R = (1+\alpha)U - \Gamma U \Gamma$ with elements $r_{ij} = (1-\gamma_i\gamma_j+\alpha)n_i^{-1}n_j s_{ij}$, Gerschgorin's Theorem applied to the matrix $R$ states that all eigenvalues $\lambda_l$ of $R$ are lying in at least one of the circular disks with centers $r_{ii}$ and radii $\sum_{j(j\neq i)} |r_{ij}|$:

$$|\lambda_l - r_{ii}| \leq \sum_{j(j\neq i)} |r_{ij}|, \quad \forall l, i$$

or $|\lambda_l - (1-\gamma_i^2+\alpha)s_{ii}| \leq n_i^{-1} \sum_{j(j\neq i)} |(1-\gamma_i\gamma_j+\alpha)n_j s_{ij}|$. Because $\gamma_i \in [0,1]$ and in order to have $R > 0$:

$$\alpha n_i s_{ii} > (1+\alpha) \sum_{j(j\neq i)} |n_j s_{ij}|, \quad \forall i.$$

Defining $Q = SN = P^T P N$ this corresponds to

$$q_{ii} > \delta_Q \sum_{j(j\neq i)} |q_{ij}|, \quad \forall i$$

with $\delta_Q = (1+\alpha)/\alpha$.
Hence under that condition

$$\begin{aligned} V_{k+1} &= \|P_1 \Gamma_1 q_k\|_2 \\ &\leq (1+\alpha_1)^{1/2} \|P_1 q_k\|_2 \\ &\leq (1+\alpha_1)^{1/2} \|P_1 V_1 P_2^{-1}\|_2 \|P_2 \Gamma_2 r_k\|_2 \end{aligned}$$

where $r_k = \|P_2 \Gamma_2 V_2 \ldots \Gamma_q V_q p_k\|_2$. The same procedure can now be repeated for $\|P_2 \Gamma_2 r_k\|_2$ as was done for $\|P_1 \Gamma_1 q_k\|_2$ etc. Hence finally

$$V_{k+1} \leq \prod_{i=1(\mathrm{mod} q)}^{q} (1+\alpha_i)^{1/2} \|P_i V_i P_{i+1}^{-1}\|_2 V_k$$

with $P_{q+1} = P_1$ and diagonal dominant matrices $Q_i = P_i^T P_i N_i$ with level of diagonal dominance equal to $\delta_{Q_i} = (1+\alpha_i)/\alpha_i$.

Defining $\beta_P = \prod_{i=1(\text{mod}q)}^{q} \|P_i V_i P_{i+1}^{-1}\|_2$ and $c_\alpha = \prod_{i=1}^{q}(1+\alpha_i)^{1/2}$, global asymptotic stability of the NL$_q$ system is obtained for

$$\beta_P < 1/c_\alpha$$

or if

$$\max_i\{\|P_i V_i P_{i+1}^{-1}\|_2 : i = 1, ..., q \ (\text{mod}q)\} \leq \beta_P^{1/q} < 1/c_\alpha$$

or if

$$\|P_{tot} V_{tot} P_{tot}^{-1}\|_2^q \leq 1/c_\alpha$$

where $P_{tot} = \text{blockdiag}\{P_2, P_3, ..., P_q, P_1\}$.

## Proof of Lemma 5.3

By definition $\|X_Q\|_\infty = \max_i \sum_j |x_{Q_{ij}}| = \max_i \sum_j d_{Q_i}^{-1} |h_{Q_{ij}}|$. Assuming $d_{Q_i} > 0$ the condition $\|X_Q\|_\infty < 1/\delta_Q$ corresponds to $\max_i \sum_j d_{Q_i}^{-1}|h_{Q_{ij}}| < 1/\delta_Q$ or $d_{Q_i} > \delta_Q \sum_{j(j\neq i)} |h_{Q_{ij}}|$ $(\forall i)$. Because $h_{Q_{ii}} = 0$ by definition, the latter means $q_{ii} > \delta_Q \sum_{j(j\neq i)} |q_{ij}|$ $(\forall i)$.

## Proof of Theorem 5.3

Propose again a Lyapunov function $V(p) = \|P_1 p\|_2$ (positive, radially unbounded and $V(0) = 0$) with $P_1 \in \mathbb{R}^{n_p \times n_p}$. Following a similar procedure as in the proof of Theorem 5.2 one obtains

$$\begin{aligned} V_{k+1} &= \|P_1 \Gamma_1 P_1^{-1} P_1 V_1 ... \Gamma_q P_q^{-1} P_q V_q P_1^{-1} P_1 p_k\|_2 \\ &\leq \|P_1 \Gamma_1 P_1^{-1}\|_2 \|P_1 V_1 P_2^{-1}\|_2 \|P_2 \Gamma_2 P_2^{-1}\|_2 \|P_2 V_2 P_3^{-1}\|_2 ... \\ &\quad \|P_q \Gamma_q P_q^{-1}\|_2 \|P_q V_q P_1^{-1}\|_2 V_k \\ &\leq \prod_{i=1}^{q} \kappa(P_i)^2 \|P_1 V_1 P_2^{-1}\|_2 \|P_2 V_2 P_3^{-1}\|_2 ... \|P_q V_q P_1^{-1}\|_2 V_k. \end{aligned}$$

because $\|\Gamma_i\|_2 \leq 1$.

## Proof of Theorem 5.5

Defining the additional state variable $\xi_k = f(C x_k)$ the closed loop system

is of the type $NL_1$ with $p_k = [x_k; \xi_k]$ in $p_{k+1} = \Gamma(p_k)V_{tot}p_k$ where $V_{tot} = [A \ -B; CA \ -CB]$ and $\Gamma = \text{diag}\{I, \Gamma_C\}$. Here $\|\Gamma_C\| < K$ because $\gamma_{C_i} = f(\varphi_k)/\varphi_k \leq K$ $(\varphi_k \neq 0)$ and the nonlinearity $f(.)$ belongs to a sector $[0, K]$ $(K \geq 1)$ and $\xi_{k+1} = f(\varphi_k) = \Gamma_C(\varphi_k)\varphi_k$ where by definition $\varphi_k = CAx_k - CB\xi_k$.
The case of diagonal scaling is then easily proven by proposing the Lyapunov function $V(p) = \|Dp\|_2$ where $D$ is a diagonal matrix. Hence

$$V_{k+1} = \|D\Gamma(p_k)V_{tot}p_k\|_2 \leq K\|DV_{tot}D^{-1}\|_2 V_k$$

which proves the Theorem.
For the second case a Lyapunov function $V(p) = \|Pp\|_2$ is proposed. Like in Theorem 5.2 a condition on $P$ and $\alpha$ is sought such that the following holds

$$V_{k+1} = \|P\Gamma V_{tot}p_k\|_2 \leq (1+\alpha)^{1/2}\|PV_{tot}P^{-1}\|_2 V_k.$$

Gerschgorin's Theorem is applied to $R = (1+\alpha)U - \Gamma U \Gamma$, with $U = N^{-1}SN$ $S = P^T P$, but now $\|\Gamma\| \leq K$ $(\gamma_i \in [0, K])$. Hence

$$(1 - K^2 + \alpha)n_i s_{ii} \geq (1+\alpha) \sum_{j(j \neq i)} |n_j s_{ij}|.$$

Defining $Q = SN$, then

$$q_{ii} \geq \delta_Q \sum_{j(j\neq i)} |q_{ij}|, \qquad \delta_Q = (1+\alpha)/(1 - K^2 + \alpha)$$

and $(1+\alpha)^{1/2} = K(\delta_Q/(\delta_Q - 1))^{1/2}$.
For the third case again a Lyapunov function $V(p) = \|Pp\|_2$ is proposed. It follows immediately then that

$$V_{k+1} \leq \|P\Gamma(p_k)P^{-1}\, PV_{tot}P^{-1}\, Pp_k\|_2 \leq K\, \|P\|_2\|P^{-1}\|_2\|PV_{tot}P^{-1}\|_2 V_k$$

which completes the proof.

## Proof of Theorem 5.6

Defining $D_{S_1} = \text{diag}\{D_1, I_{n_w}\}$ it can be shown that

$$\begin{aligned} \|D_{S_1} \begin{bmatrix} p_{k+1} \\ e_k^{ext} \end{bmatrix}\|_2^2 &= \|D_1 p_{k+1}\|_2^2 + \|e_k\|_2^2 \\ &\leq \|D_{S_1} R_1 D_2^{-1}\|_2^2 \|D_2 R_2 D_3^{-1}\|_2^2 \ldots \\ &\quad \|D_q R_q D_{S_1}^{-1}\|_2^2 (\|D_1 p_k\|_2^2 + \|w_k\|_2^2) \end{aligned}$$

according to the proof of Theorem 5.1, because $\|\Omega_i\|_2 \leq 1$ and with $D_i$ ($i = 2, ..., q$) diagonal with nonzero diagonal elements.
Defining $\beta_D = \|D_{S_1} R_1 D_2^{-1}\|_2 \prod_{i=2}^{q-1} \|D_i R_i D_{i+1}^{-1}\|_2 \|D_q R_q D_{S_1}^{-1}\|_2$ it follows from the proof in Theorem 5.1 that $\beta_D < 1$ is satisfied if $\|D_{tot} R_{tot} D_{tot}^{-1}\|_2^q < 1$ where $D_{tot} = \text{diag}\{D_2, D_3, ..., D_q, D_{S_1}\}$.
Defining $r_k = \|D_1 p_k\|_2$, the following holds, provided that $\beta_D < 1$:

$$\begin{array}{rcl} r_{k+1}^2 + \|e_k\|_2^2 & \leq & \beta_D^2 r_k^2 + \beta_D^2 \|w_k\|_2^2 \\ \Rightarrow \quad \sum_{k=0}^{N-1} r_{k+1}^2 + \sum_{k=0}^{N-1} \|e_k\|_2^2 & \leq & \beta_D^2 \sum_{k=0}^{N-1} r_k^2 + \beta_D^2 \sum_{k=0}^{N-1} \|w_k\|_2^2 \\ \Rightarrow \quad r_N^2 + (1 - \beta_D^2) \sum_{k=0}^{N-1} r_k^2 + \sum_{k=0}^{N-1} \|e_k\|_2^2 & \leq & r_0^2 + \beta_D^2 \sum_{k=0}^{N-1} \|w_k\|_2^2 \\ \Rightarrow \quad \sum_{k=0}^{N-1} \|e_k\|_2^2 & \leq & r_0^2 + \beta_D^2 \sum_{k=0}^{N-1} \|w_k\|_2^2. \end{array}$$

Furthermore for $N \to \infty$: $(1 - \beta_D^2)\|r\|_2^2 + \|e\|_2^2 \leq r_0^2 + \beta_D^2 \|w\|_2^2$. Defining constants $c_1 = \overline{\sigma}_{D_1}^2$, $c_2 = \underline{\sigma}_{D_1}^2$ one obtains the result.

## Proof of Theorem 5.7

Defining $P_{S_1}$ = blockdiag$\{P_1, I_{n_w}\}$ and according to the proofs in Theorem 5.1 and 5.6

$$\begin{array}{rcl} \|P_{S_1} \begin{bmatrix} p_{k+1} \\ e_k^{ext} \end{bmatrix}\|_2^2 & = & \|P_1 p_{k+1}\|_2^2 + \|e_k\|_2^2 \\ & \leq & \prod_{i=1}^q (1+\alpha_i) \|P_{S_1} R_1 P_2^{-1}\|_2^2 ... \|P_i R_i P_{i+1}^{-1}\|_2^2 ... \\ & & \|P_q R_q P_{S_1}^{-1}\|_2^2 (\|P_1 p_k\|_2^2 + \|w_k\|_2^2) \end{array}$$

with diagonal dominant matrices $Q_i = P_i^T P_i N_i$, $\delta_{Q_i} = (1+\alpha_i)/\alpha_i$. Defining $c_\alpha = \prod_{i=1}^q (1+\alpha_i)^{1/2}$, $\beta_P = \|P_{S_1} R_1 P_2^{-1}\|_2 \prod_{i=2}^{q-1} \|P_i R_i P_{i+1}^{-1}\|_2 \|P_q R_q P_{S_1}^{-1}\|_2$ and according to Theorem 5.2, $c_\alpha \beta_P < 1$ holds if $\|P_{tot} R_{tot} P_{tot}^{-1}\|_2^q < 1/c_\alpha$ where $P_{tot} = \text{blockdiag}\{P_2, P_3, ..., P_q, P_{S_1}\}$. Putting $r_k = \|P_1 p_k\|_2$ then

$$\begin{array}{rcl} r_{k+1}^2 + \|e_k\|_2^2 & \leq & c_\alpha^2 \beta_P^2 r_k^2 + c_\alpha^2 \beta_P^2 \|w_k\|_2^2 \\ \Rightarrow \quad r_N^2 + (1 - c_\alpha^2 \beta_P^2) \sum_{k=0}^{N-1} r_k^2 + \sum_{k=0}^{N-1} \|e_k\|_2^2 & \leq & r_0^2 + c_\alpha^2 \beta_P^2 \sum_{k=0}^{N-1} \|w_k\|_2^2 \\ \Rightarrow \quad \sum_{k=0}^{N-1} \|e_k\|_2^2 & \leq & r_0^2 + c_\alpha^2 \beta_P^2 \sum_{k=0}^{N-1} \|w_k\|_2^2. \end{array}$$

Defining constants $c_1 = \overline{\sigma}_{P_1}^2$, $c_2 = \underline{\sigma}_{P_1}^2$ and letting $N \to \infty$ one obtains the result.

## Proof of Theorem 5.8

Defining again $P_{S_1} = \text{blockdiag}\{P_1, I_{n_w}\}$ one obtains

$$\begin{array}{rcl} \|P_{S_1} \left[ \begin{array}{c} p_{k+1} \\ e_k^{ext} \end{array} \right] \|_2^2 & = & \|P_1 p_{k+1}\|_2^2 + \|e_k\|_2^2 \\ & \leq & \kappa(P_{S_1})^2 \prod_{i=2}^q \kappa(P_i)^2 \, \|P_{S_1} R_1 P_2^{-1}\|_2^2 ... \|P_i R_i P_{i+1}^{-1}\|_2^2 ... \\ & & \|P_q R_q P_{S_1}^{-1}\|_2^2 (\|P_1 p_k\|_2^2 + \|w_k\|_2^2) \end{array}$$

and $\kappa(P_{S_1}) = \kappa(P_1)$.

## Proof of Lemma 5.5

This follows immediately from the proof of Theorem 5.6, 5.7 and 5.8. Take e.g. the diagonal dominance case: if $c_\alpha \beta_P < 1$ then $r_{k+1}^2 + \|e_k\|_2^2 \leq c_\alpha^2 \beta_P^2 r_k^2 + c_\alpha^2 \beta_P^2 \|w_k\|_2^2 \leq r_k^2 + c_\alpha^2 \beta_P^2 \|w_k\|_2^2$ or $r_{k+1}^2 - r_k^2 \leq c_\alpha^2 \beta_P^2 \|w_k\|_2^2 - \|e_k\|_2^2$ which proves the Lemma.

## Proof of Theorem 5.9

Consider the perturbed $NL_q$ written as an $NL_{q+1}$ system. Defining $P_{S_1} = \text{blockdiag}\ \{P_1, I, D_1\}$ and $\Upsilon_{q+1} = \text{diag}\{I, I, \Delta_x\}$ one obtains like in the proof of Theorem 5.7:

$$\begin{array}{rcl} \|P_{S_1} \left[ \begin{array}{c} p_{k+1} \\ e_k^{ext} \\ s_k \end{array} \right] \|_2^2 & = & \|P_1 p_{k+1}\|_2^2 + \|e_k\|_2^2 + \|D_1 s_k\|_2^2 \\ & \leq & \prod_{i=1}^q (1+\alpha_i) \|P_{S_1} T_1 P_2^{-1}\|_2^2 ... \|P_i T_i P_{i+1}^{-1}\|_2^2 \\ & & ... \|P_q T_q P_{S_1}^{-1} P_{S_1} \Upsilon_{q+1} \left[ \begin{array}{c} p_k \\ w_k \\ s_k \end{array} \right] \|_2^2 \\ & \leq & \prod_{i=1}^q (1+\alpha_i) \|P_{S_1} T_1 P_2^{-1}\|_2^2 ... \|P_i T_i P_{i+1}^{-1}\|_2^2 ... \|P_q T_q P_{S_1}^{-1}\|_2^2 \\ & & (\|P_1 p_k\|_2^2 + \|w_k\|_2^2 + \|D_1 s_k\|_2^2) \end{array}$$

because $\Upsilon_{q+1}$ commutes with $P_{S_1}$ and $T_{q+1} = I$. The matrices $Q_i = P_i^T P_i N_i$ are again diagonal dominant. Defining $c_\alpha = \prod_{i=1}^q (1+\alpha_i)^{1/2}$, $\beta_P = \|P_{S_1} T_1 P_2^{-1}\|_2 \prod_{i=2}^{q-1} \|P_i T_i P_{i+1}^{-1}\|_2 \|P_q T_q P_{S_1}^{-1}\|_2$ and according to Theorem 5.7, $c_\alpha \beta_P < 1$ holds if $\|P_{tot} T_{tot} P_{tot}^{-1}\|_2^q < 1/c_\alpha$ where $P_{tot} = \text{blockdiag}\ \{P_2, P_3,$

..., $P_q$, $P_{S_1}$}. Putting $r_k = \|P_1 p_k\|_2$ then

$$\begin{aligned} r_{k+1}^2 + \|e_k\|_2^2 + \|D_1 s_k\|_2^2 &\leq c_\alpha^2 \beta_P^2 (r_k^2 + \|w_k\|_2^2 + \|D_1 s_k\|_2^2) \\ &\leq c_\alpha^2 \beta_P^2 r_k^2 + c_\alpha^2 \beta_P^2 \|w_k\|_2^2 + \|D_1 s_k\|_2^2 \\ \Rightarrow \quad r_{k+1}^2 + \|e_k\|_2^2 &\leq r_k^2 + \|w_k\|_2^2 \end{aligned}$$

because $c_\alpha \beta_P < 1$.
The second case follows immediately from

$$\begin{aligned} \|P_{S_1} \begin{bmatrix} p_{k+1} \\ e_k^{ext} \\ s_k \end{bmatrix}\|_2^2 \leq\ & \kappa(P_{S_1})^2 \prod_{i=2}^{q} \kappa(P_i)^2 \, \|P_{S_1} T_1 P_2^{-1}\|_2^2 ... \|P_i T_i P_{i+1}^{-1}\|_2^2 \\ & ... \|P_q T_q P_{S_1}^{-1}\|_2^2 \|P_{S_1} \begin{bmatrix} p_k \\ w_k \\ s_k \end{bmatrix}\|_2^2 \end{aligned}$$

# Bibliography


Albertini F., Sontag E.D. (1993). For neural networks, function determines form, *Neural Networks*, Vol.6, pp.975-990.

Alspach D.L., Sorenson H.W. (1972). Nonlinear Bayesian estimation using Gaussian sum approximations, *IEEE Transactions on Automatic Control*, Vol.17, No.4, pp.439-448.

Amari S.-i., Kurata K., Nagaoka H. (1992). Information geometry of Boltzmann machines, *IEEE Transactions on Neural Networks*, Vol.3, No.2, pp.260-271.

Anderson B.D.O., Moore J.B. (1979). *Optimal filtering*, Prentice-Hall.

Arena P., Baglio P., Fortuna F., Manganaro G. (1994). Generation of $n$-double scrolls via cellular neural networks, *submitted to International Journal of Circuit Theory and Applications (Special Issue on Cellular Neural Networks)*.

Arnold L. (1974). *Stochastic differential equations: theory and applications*, John Wiley & Sons.

Åström K.J. (1970). *Introduction to stochastic control theory*, Academic Press.

Åström K.J., Eykhoff P. (1971). System identification - a survey, *Automatica*, Vol.7, pp.123-162.

Åström K.J., Wittenmark B. (1984). *Computer-controlled systems, theory and design*, Englewood Cliffs NJ: Prentice-Hall.

Barron A.R. (1993). Universal approximation bounds for superposition of a sigmoidal function, *IEEE Transactions on Information Theory*, Vol.39, No.3, pp.930-945.

Barron A.R. (1993). When do neural nets avoid the curse of dimensionality ?, *NATO advanced study institute - From statistics to neural networks*, June 21 - July 2, Les Arcs, France.

Barto A.G., Sutton R.S., Anderson C.W. (1983). Neuronlike adaptive elements that can solve difficult learning control problems, *IEEE Transactions on Systems, Man, and Cybernetics*, Vol. SMC-13, No.5, pp.834-846.

Bengio Y., Simard P., Frasconi P. (1994). Learning long-term dependencies with gradient descent is difficult, *IEEE Transactions on Neural Networks*, Vol.5, No.2, pp.157-166.

Billings S.A., Voon W.S.F. (1986). Correlation based model validity tests for non-linear models, *International Journal of Control*, Vol.44, No.1, pp.235-244.

Billings S.A., Chen S. (1989). Extended model set, global data and threshold model identification of severely non-linear systems, *International Journal of Control*, Vol.50, No.5, pp.1897-1923.

Billings S.A., Jamaluddin H., Chen S. (1992). Properties of neural networks with applications to modelling non-linear dynamical systems, *International Journal of Control*, Vol.55, No.1, pp.193-224.

Boyd S., Yang Q. (1989). Structured and simultaneous Lyapunov functions for system stability problems, *International Journal of Control*, Vol.49, No.6, pp.2215-2240.

Boyd S., Barratt C., Norman S. (1990). Linear controller design: limits of performance via convex optimization, *Proceedings of the IEEE*, Vol.78, No.3, pp.529-574.

Boyd S., Barratt C. (1991). *Linear controller design, limits of performance*, Prentice-Hall.

Boyd S., El Ghaoui L. (1993). Method of centers for minimizing generalized eigenvalues, *Linear Algebra and Applications, special issue on Numerical Linear Algebra Methods in Control, Signals and Systems*, 188, July.

Boyd S., El Ghaoui L., Feron E., Balakrishnan V. (1994). *Linear matrix inequalities in system and control theory*, SIAM (Studies in Applied Mathematics), Vol.15.

Brockett R.W. (1966). The status of stability theory for deterministic systems, *IEEE Transactions on Automatic Control*, pp.597-605, July.

Broomhead D.S., Lowe D. (1988). Multivariable functional interpolation and adaptive networks, *Complex Systems*, 2, pp.321-355.

Bryson A.E., Ho Y.C. (1969). *Applied optimal control*, Waltham, MA: Blaisdel.

Byrnes C.I., Lin W. (1994). Losslessness, feedback equivalence and the global stabilization of discrete-time nonlinear systems, *IEEE Transactions on Automatic Control*, Vol.39, No.1, pp.83-98.

Casalino G., Parisini T., Zoppoli R. (1993). A neural optimal controller for a real space robot, *World Congress on Neural Networks '93, Portland*, Vol.III, pp.208-211.

Casdagli M. (1989). Nonlinear prediction of chaotic time series, *Physica D*, 35, pp.335-356.

Casdagli M., Eubank S., Farmer J.D., Gibson J. (1991). State space reconstruction in the presence of noise, *Physica D*, 51, pp.52-98.

Chakraborty K., Mehrotra K., Mohan C.K., Ranka S. (1992). Forecasting the behaviour of multivariate time series using neural networks, *Neural Networks*, Vol.5, pp.961-970.

Charalambous C. (1992). Conjugate gradient algorithm for efficient training of artificial neural networks, *IEE Proceedings-G*, Vol.139, No.3, pp.301-310.

Chen S., Billings S.A., Luo W. (1989). Orthogonal least squares methods and their application to non-linear system identification, *International Journal of Control*, Vol.50, No.5, pp.1873-1896.

Chen S., Billings S., Grant P. (1990a). Nonlinear system identification using neural networks, *International Journal of Control*, Vol.51, No.6, pp.1191-1214.

Chen S., Cowan C., Billings S., Grant P. (1990b). Parallel recursive prediction error algorithm for training layered neural networks, *International Journal of Control*, Vol.51, No.6, pp.1215-1228.

Chen S., Cowan C., Grant P. (1991). Orthogonal least squares learning algorithm for radial basis function networks, *IEEE Transactions on Neural Networks*, Vol.2, No.2, pp.302-309.

Chen S., Billings S.A. (1992). Neural networks for nonlinear dynamic system modelling and identification, *International Journal of Control*, Vol.56, No.2, pp.319-346.

Chen S., Billings S.A., Grant M. (1992b). Recursive hybrid algorithm for non-linear system identification using radial basis function networks, *International Journal of Control*, Vol.55, No.5, pp.1051-1070.

Cheney E.W. (1966). *Introduction to approximation theory*, McGraw-Hill.

Chua L.O., Yang L. (1988). Cellular neural networks: Theory & Applications, *IEEE Transactions on Circuits and Systems*, Vol. 35, No.10, pp.1257-1290.

Chua L.O., Roska T. (1993). The CNN Paradigm, *IEEE Transactions on Circuits and Systems-I*, Vol.40, No.3, pp.147-156.

Chua L.O. (1994). Chua's circuit 10 years later, *International Journal of Circuit Theory and Applications*, Vol.22, pp.279-305.

Chua L.O., Komuro M., Matsumoto T. (1986). The Double Scroll Family, *IEEE Transactions on Circuits and Systems-I*, Vol. CAS-33, No.11, pp.1072-1118.

Chua L.O. (1992). The genesis of Chua's Circuit, *Archiv für Elektrotechnik und Ubertragungtechnik*, Vol.46, No.4, pp.250-257.

Chua L.O. (1993). Global unfolding of Chua's Circuit, *IEICE Trans. Fundamentals*, Vol. E76-A, No.5, pp.704-734.

Chua L.O., Wu C.W., Huang A., Zhong G-Q (1993). A universal circuit for studying and generating chaos - Part I: routes to chaos & Part II: strange attractors, *IEEE Transactions on Circuits and Systems-I*, Vol.40, No.10, pp.732-744 & pp.745-761.

Cuthbert T.R. (1987). *Optimization using personal computers with applications to electrical networks*, John Wiley.

Cybenko G. (1989). Approximations by superpositions of a sigmoidal function, *Mathematics of Control, Signals and Systems,* 2, pp.183-192.

Dahleh M.A., Khammash M.H. (1993). Controller design for plants with structured uncertainty, *Automatica, Special Issue on Robust Control*, Vol.29, No.1, pp.37-56.

De Moor B. (1988). *Mathematical concepts and techniques for modelling of static and dynamic systems*, PhD Thesis, Department of Electrical Engineering, Katholieke Universiteit Leuven, Belgium.

De Moor B., Van Overschee P., Suykens J. (1991). Subspace algorithms for system identification and stochastic realization, *Recent Advances in Mathematical Theory of Systems, Control, Networks and Signal Processing, Proc. of the International Symposium MTNS-91, Kobe, Japan*, June 17-21, MITA Press, pp.589-595.

Desoer C.A., Vidyasagar M. (1975). *Feedback systems: input-output properties*, New York: Academic Press.

Ditto W.L., Pecora L.M. (1993). Mastering chaos, *Scientific American*, August, pp.62-68.

Doob J.L. (1953). *Stochastic processes*, John Wiley & Sons.

Doyle J. (1982). Analysis of feedback systems with structured uncertainties, *IEE Proceedings-D*, Vol.129, No.6, pp.243-250.

Doyle J., Packard A., Zhou K. (1991). Review of LFTs, LMIs, and $\mu$, *Proceedings of the 30th Conference on Decision and Control*, Brighton, England, pp.1227-1232, Dec.

Ecker J.G., Kupferschmid M. (1985). A computational comparison of the ellipsoid algorithm with several nonlinear programming algorithms, *SIAM Journal on Control and Optimization*, Vol.23, No.5, pp.657-674.

Fletcher R. (1987). *Practical methods of optimization*, second edition, Chichester and New York: John Wiley and Sons.

Franklin G.F., Powell J.D., Workman M.L. (1990). *Digital control of dynamic systems*, Reading MA: Addison-Wesley.

Freeman W.J. (1993). Chaotic dynamics in neural pattern recognition, *NATO Advanced Study Institute - From Statistics to Neural Networks*, June 21-July 2, Les Arcs, France.

Funahashi K.-I. (1989). On the Approximate Realization of continuous Mappings by Neural Networks, *Neural Networks*, Vol.2, pp.183-192.

Gahinet, P., Nemirovskii A. (1993). General-Purpose LMI Solvers with Benchmarks, *Proceedings Conference on Decision and Control*, pp.3162-3165.

Garcia C.E., Prett D.M., Morari M. (1989). Model predictive control: theory and practice - a survey, *Automatica*, Vol.25, No.3, pp.335-348.

Gelfand S.B., Mitter S.K. (1991). Recursive stochastic algorithms for global optimization in $\mathbb{R}^d$, *SIAM Journal on Control and Optimization*, Vol.29, No.5, pp.999-1018.

Gelfand S.B., Mitter S.K. (1993). Metropolis-type annealing algorithms for global optimization in $\mathbb{R}^d$, *SIAM Journal on Control and Optimization*, Vol.31, No.1, pp.111-131.

Gihman I.I., Skorohod A.V. (1979). *The theory of stochastic processes I,II,III*, New York: Springer-Verlag.

Gill P.E., Murray W., Wright M.H. (1981). *Practical Optimization*, London: Academic Press.

Girosi F., Poggio T. (1989). Representation properties of networks: Kolmogorov's theorem is irrelevant, *Neural Computation*, 1:465-469.

Goldberg D.E. (1989). *Genetic algorithms in search, optimization and machine learning*, Addison-Wesley, Reading, Mass.

Golub G.H., Van Loan C.F. (1989). *Matrix Computations*, Baltimore MD: Johns Hopkins University Press.

Goodwin G., Sin K. (1984). *Adaptive filtering, prediction and control*, Prentice-Hall.

Grossberg S. (1988). Nonlinear neural networks: principles, mechanisms, and architectures, *Neural Networks*, Vol.1, pp.17-61.

Gu D.-W., Tsai M.C., O'Young S.D., Postlethwaite I. (1989). State-space formulae for discrete-time $H_\infty$ optimization, *International Journal of Control*, Vol.49, 1683-1723.

Guckenheimer J., Holmes P. (1983). *Nonlinear Oscillations, Dynamical Systems, and Bifurcations of Vector Fields*, New York: Springer Verlag.

Guzelis C., Chua L.O. (1993). Stability analysis of generalized cellular neural networks, *International Journal of Circuit Theory and Applications*, Vol.21, pp.1-33.

Hammerstrom D. (1993). Working with neural networks, *IEEE Spectrum*, July, pp.46-53.

Harrer H., Nossek J.A. (1992). Discrete time cellular neural networks, *International Journal of Circuit Theory and Applications*, Vol.20, pp.453-467.

Harrer H. (1993). Multiple layer discrete-time cellular neural networks using time-variant templates, *IEEE Transactions on Circuits and Systems-II*, Vol. 40, No.3, pp.191-199.

Hebb D.D. (1949). *The organization of behaviour*, John Wiley & Sons, New York.

Hecht-Nielsen R. (1987). Kolmogorov's mapping neural network existence theorem, *Proc. IEEE International Conference on Neural Networks.*

Hecht-Nielsen R. (1988a). Neurocomputing: picking the human brain, *IEEE Spectrum*, 25(3), pp.36-41.

Hecht-Nielsen R. (1988b). Neurocomputer applications, in *NATO ASI Series, Vol.F 41, Neural computers*, Eds. R. Eckmiller, Ch. v.d. Malsburg, Springer-Verlag.

Hestenes M. (1990). *Conjugate direction methods in optimization*, New York: Springer-Verlag.

Hill D.J., Moylan P.J. (1980). Connections between finite-gain and asymptotic stability, *IEEE Transactions on Automatic Control*, Vol.AC-25, No.5, pp.931-936.

Hill D.J., Moylan P.J. (1976). The stability of nonlinear dissipative systems, *IEEE Transactions on Automatic Control*, Vol.AC-21, pp.708-711.

Hinton G.E. (1989). Connectionist Learning Procedures, *Artificial Intelligence*, 40, pp.185-234.

Hirsch M.W., Smale S. (1974). *Differential Equations, Dynamical Systems, and Linear Algebra*, New York: Academic Press.

Hopfield J.J. (1982). Neural networks and physical systems with emergent collective computational abilities, *Proceedings of the National Academy of Sciences of the USA*, 79, April, pp.2554-2558.

Horn R.A., Johnson C.R. (1985). *Matrix Analysis*, Cambridge University Press.

Hornik K., Stinchcombe M., White H. (1989). Multilayer feedforward networks are universal approximators, *Neural Networks*, Vol.2, pp.359-366.

Hornik K. (1991). Approximation capabilities of multilayer feedforward networks, *Neural Networks*, Vol.4, pp.251-257.

Hunt K.J., Sbarbaro D. (1991). Neural networks for nonlinear internal model control, *IEE Proceedings-D*, Vol.138, No.5, pp.431-438.

Hunt K.J., Sbarbaro D., Zbikowski R., Gawthrop P.J. (1992). Neural networks for control systems - a survey, *Automatica*, Vol. 28., No. 6, pp.1083-1112.

Iglesias P.A., Glover K. (1991). State-space approach to discrete-time $H_\infty$ control, *International Journal of Control*, Vol.54, 1031-1073.

Ikeda N., Watanabe S. (1981). *Stochastic differential equations and diffusion processes*, North-Holland / Kodansha.

Isidori A. (1985). *Nonlinear control systems: an introduction*, Springer, Berlin-New York.

Johansson E.M., Dowla F.U., Goodman D.M. (1992). Backpropagation learning for multilayer feed-forward neural networks using the conjugate gradient method, *International Journal of Neural Systems*, Vol.2, No.4, pp.291-301.

Jury E.I. (1974). *Inners and stability of dynamic systems*, John Wiley & Sons.

Kaizer A.J.M. (1987). Modeling of the nonlinear response of an electrodynamic loudspeaker by a Volterra series expansion, *J. Audio Eng. Soc.*, Vol.35, No.6, 421-433.

Karmarkar N. (1984). A new polynomial-time algorithm for linear programming, *Combinatorica*, 4, pp.373-395.

Kaszkurewicz E., Bhaya A. (1993). Robust stability and diagonal Liapunov functions, *SIAM Journal on Matrix Analysis and Applications*, Vol.14, No.2, pp.508-520.

Kato T. (1976). *Perturbation theory for linear operators*, Springer-Verlag.

Kirkpatrick S., Gelatt C.D., Vecchi M. (1983). Optimization by simulated annealing, *Science*, 220, pp.621-680.

Klippel W. (1990). Dynamic measurement and interpretation of the nonlinear parameters of electrodynamic loudpeakers, *J. Audio Eng. Soc.*, Vol.38, No.12, pp.944-955.

Klippel W. (1992a). Nonlinear large-signal behavior of electrodynamic loudspeakers at low frequencies, *J. Audio Eng. Soc.*, Vol.40, No.6, pp.483-496.

Klippel W. (1992b). The mirror filter-a new basis for reducing nonlinear distortion and equalizing response in woofer systems, *J. Audio Eng. Soc.*, Vol.40, No.9, 675-691.

Kohonen T. (1977). *Associative memory - a systemtheoretical approach*, Springer-Verlag, Berlin.

Kosmatopoulos E.B., Christodoulou M.A. (1994). The Boltzmann g-RHONN: a learning machine for estimating unknown probability distributions, *Neural Networks*, Vol.7, No.2, pp.271-278.

Kůrková V. (1991). Kolmogorov's theorem is relevant, *Neural Computation*, 3, pp.617-622.

Kůrková V. (1992). Kolmogorov's theorem and multilayer neural networks, *Neural Networks*, 5, pp.501-506.

Kushner H.J. (1987). Asymptotic global behavior for stochastic approximation and diffusions with slowly decreasing noise effects: global minimization via Monte Carlo, *SIAM Journal on Applied Mathematics*, 47, pp.169-185.

Larcombe P. (1991). On the control of a two dimensional multi-link inverted pendulum: co-ordinate system suitable for dynamic formulation, *Proceedings of the 30th Conference on Decision and Control*, pp.136-141, Brighton, England.

Le Cun Y. (1988). A theoretical framework for back-propagation, *Proceedings of the 1988 Connectionist Models Summer School*, San Mateo, CA: Morgan Kaufmann, (Eds. D. Touretzky, G. Hinton, T. Sejnowski).

Leontaritis I.J., Billings S.A. (1985). Input-output parametric models for nonlinear systems, Part I: deterministic non-linear systems & Part II: stochastic non-linear systems, *International Journal of Control*, Vol.41, No.2, pp.303-328 & pp.329-344.

Leshno M., Lin V.Y., Pinkus A., Schocken S. (1993). Multilayer feedforward networks with a nonpolynomial activation function can approximate any function, *Neural Networks*, Vol.6, pp.861-867.

Lippmann R.P. (1987). Introduction to computing with neural nets, *IEEE ASSP Magazine*, April, pp.4-22.

Liu C.-C., Chen F.-C. (1993). Adaptive control of non-linear continuous-time systems using neural networks - general relative degree and MIMO cases, *International Journal of Control*, Vol.58, No.2, pp.317-335.

Liu D., Michel A.N. (1992). Asymptotic stability of discrete time systems with saturation nonlinearities with applications to digital filters, *IEEE Transactions on Circuits and Systems-I*, Vol.39, No.10, pp.798-807.

Ljung L. (1977). Analysis of recursive stochastic algorithms, *IEEE Transactions on Automatic Control*, Vol.AC-22, No.4, pp.551-575.

Ljung L. (1979). Asymptotic behavior of the extended Kalman filter as a parameter estimator for linear systems, *IEEE Transactions on Automatic Control*, Vol.AC-24, No.1, pp.36-50.

Ljung L. (1981). Analysis of a general recursive prediction error identification algorithm, *Automatica*, Vol.17, No.1, pp.89-99.

Ljung L. (1987). *System Identification: Theory for the User*, Prentice-Hall.

Luenberger D.G. (1979). *Introduction to dynamic systems: theory, models and applications*, John Wiley & Sons.

Maciejowski J.M. (1989). *Multivariable feedback design*, Addison-Wesley.

Madan R.N. (1993). Guest Editor: 'Special Issue on Chua's Circuit: A Paradigm for Chaos', *Journal of Circuits, Systems and Computers, Part 1*, Vol.3, No.1, March 1993 and *Part 2*, Vol.3, No.2, June 1993.

Mayer-Kress G., Choi I., Weber N., Bargar R., Hübler A. (1993). Musical signals from Chua's circuit, *IEEE Transactions on Circuits and Systems-II*, Vol.40, No.10, pp.688-695.

Miller W.T., Sutton R.S., Werbos P.J. (1990). *Neural networks for control*, Cambridge, MA: MIT Press.

Møller M.F. (1993). A scaled conjugate gradient algorithm for fast supervised learning, *Neural Networks*, Vol.6, pp.525-533.

Moonen M., De Moor B., Vandenberghe L., Vandewalle J. (1989). *On- and off-line identification of linear state-space models*, International Journal of Control, Vol.49, pp.219-232.

Narendra K.S., Taylor J.H. (1973). *Frequency domain criteria for absolute stability*, Academic Press.

Narendra K.S., Parthasarathy K. (1990). Identification and control of dynamical systems using neural networks, *IEEE Transactions on Neural Networks*, Vol.1, No.1, pp.4-27.

Narendra K.S., Parthasarathy K. (1991). Gradient methods for the optimization of dynamical systems containing neural networks, *IEEE Transactions on Neural Networks*, Vol.2, No.2, pp.252-262.

Nelson E. (1967). *Dynamical theories of Brownian motion*, New Jersey: Princeton University Press.

Nemirovskii A., Gahinet P. (1994). The Projective Method for Solving Linear Matrix Inequalities, *Proceedings American Control Conference*, pp.840-844.

Nerrand O., Roussel-Ragot P., Personnaz L., Dreyfus G. (1993). Neural networks and nonlinear adaptive filtering: unifying concepts and new algorithms, *Neural Computation*, 5, 165-199.

Nesterov Y., Nemirovskii A. (1994). *Interior point polynomial algorithms in convex programming*, SIAM (Studies in Applied Mathematics), Vol.13.

Nguyen D., Widrow B. (1990). Neural networks for self-learning control systems, *IEEE Control Systems Magazine*, 10(3), pp.18-23.

Nijmeijer H., van der Schaft A.J. (1990). *Nonlinear dynamical control systems*, Springer-Verlag, New York.

Ogorzalek M.J. (1993). Taming chaos - Part I: Synchronization & Part II: Control, *IEEE Transactions on Circuits and Systems-I*, Vol.40, No.10, pp.693-699 & pp.700-706.

Overton M.L. (1988). On minimizing the maximum eigenvalue of a symmetric matrix, *SIAM Journal on Matrix Analysis and Applications*, Vol.9, No.2, pp.256-268.

Packard A., Doyle J. (1993). The complex structured singular value, *Automatica*, Vol.29, No.1, pp.71-109.

Papoulis A. (1965). *Probability, random variables and stochastic processes*, Mc Graw-Hill.

Parisini T., Zoppoli R. (1991). Backpropagation for $N$-stage optimal control problems, *International Joint Conference on Neural Networks, Signapore '91*, pp.1518-1529.

Parisini T., Zoppoli R. (1994). Neural networks for feedback feedforward nonlinear control systems, *IEEE Transactions on Neural Networks*, Vol.5, No.3, pp.436-449.

Park D.C., El-Sharkawi M.A., Marks II R.J., Atlas L.E., Damborg M.J. (1991). Electric load forecasting using an artificial neural network, *IEEE Transactions on Power Systems*, Vol.6., No.2, pp.442-449.

Park J., Sandberg I.W. (1991). Universal approximation using Radial-Basis-Function networks, *Neural Computation*, 3, pp.246-257.

Parker T.S., Chua L.O. (1989). *Practical numerical algorithms for chaotic systems*, Springer-Verlag.

Perez-Munuzuri A.P., Perez-Munuzuri V., Perez-Villar V., Chua L.O. (1993). Spiral waves on a 2-D array of nonlinear circuits, *IEEE Transactions on Circuits and Systems I*, Vol.40, No.10, pp.872-877.

Piché S.W. (1994). Steepest descent algorithms for neural network controllers and filters, *IEEE Transactions on Neural Networks*, Vol.5, No.2, pp.198-212.

Poggio T., Girosi F. (1990). Networks for approximation and learning, *Proceedings of the IEEE*, Vol.78, No.9, pp.1481-1497.

Polak E., Mayne D. (1985). Algorithm models for nondifferentiable optimization, *SIAM Journal on Control and Optimization*, Vol.23, No.3, pp.477-491.

Polak E., Mayne D., Stimler D. (1984). Control system design via semi-infinite optimization: a review, *Proceedings of the IEEE*, Vol.72, No.12, pp.1777-1794.

Polak E., Salcudean S. (1989). On the design of linear multivariable feedback systems via constrained nondifferentiable optimization in $H^{\infty}$ spaces, *IEEE Transactions on Automatic Control*, Vol.34, No.3, pp.268-276.

Polak E., Wardi Y. (1982). Nondifferentiable optimization algorithm for designing control systems having singular value inequalities, *Automatica*, Vol.18, No. 3, pp.267-283.

Powell M. (1977). Restart procedures for the conjugate gradient method, *Mathematical Programming*, 12(2), pp.241-254.

Psaltis D., Sideris A., Yamamura A. (1988). A multilayered neural network controller, *IEEE Control Systems Magazine*, April, pp.17-21.

Reed R. (1993). Pruning algorithms - a survey, *IEEE Transactions on Neural Networks*, Vol.4, No.5, pp.740-747.

Rice J.R. (1983). *Numerical methods, software and analysis*, Mc Graw-Hill.

Ritter H., Martinetz T., Schulten K. (1992). *Neural computation and self-organizing maps: an introduction*, Addison-Wesley.

Rosenblatt R. (1959). *Principles of neurodynamics*, Spartan Books, New York.

Roska T., Chua L.O. (1993). The CNN Universal Machine: an analogic array computer, *IEEE Transactions on Circuits and Systems-II*, Vol.40, No.3, pp.163-173.

Roska T., Vandewalle J. (1993) (Eds.). *Cellular Neural Networks*, John Wiley and Sons.

Rumelhart D.E., Hinton G.E., Williams R.J. (1986). Learning representations by back-propagating errors, *Nature*, Vol.323, pp.533-536.

Saarinen S., Bramley R., Cybenko G. (1993). Ill-Conditioning in neural network training problems, *SIAM Journal on Scientific Computing*, Vol.14, No.3, pp.693-714.

Saerens M., Soquet A. (1991). Neural controller based on back-propagation algorithm, *IEE Proceedings-F*, Vol.138, No.1, pp.55-62.

Saerens M., Renders J.-M., Bersini H. (1993). Neural controllers based on backpropagation algorithm. In *IEEE Press Book on Intelligent Control: Theory and Practice*, M. M. Gupta, N. K. Sinha (Eds.), IEEE Press.

Sanner R.M., Slotine J.-J. E. (1992). Gaussian networks for direct adaptive control, *IEEE Transactions on Neural Networks*, Vol.3, No.6, pp.837-863.

Saravanan N., Duyar A., Guo T.-H., Merrill W.C. (1993). Modeling of the space shuttle main engine using feed-forward neural networks, *Proceedings of the American Control Conference*, San Francisco, California, pp.2897-2899.

Sbarbaro-Hofer D., Neumerkel D., Hunt K. (1993). Neural control of a steel rolling mill, *IEEE Control Systems*, June, pp.69-75.

Schiffmann W.H., Geffers H.W. (1993). Adaptive control of dynamic systems by back propagation networks, *Neural Networks*, Vol.6, pp.517-524.

Shah S., Palmieri F., Datum M. (1992). Optimal filtering algorithms for fast learning in feedforward neural networks, *Neural Networks*, Vol.5, pp.779-787.

Shor N.Z. (1985). *Minimization methods for non-differentiable functions*, Berlin, Heidelberg: Springer-Verlag.

Siljak D.D. (1989). Parameter space methods for robust control design, *IEEE Transactions on Automatic Control*, Vol.34, No.7, pp.674-688.

Singhal S., Wu L. (1989). Training feed-forward networks with the extended Kalman filter. *Proceedings of the IEEE International Conference on Acoustics, Speech and Signal Processing*, pp.1187-1190, Glasgow, Scotland: IEEE Press.

Sjöberg J., Ljung L. (1992). Overtraining, regularization and searching for minimum in neural networks, *4th IFAC International Symposium on adaptive systems in control and signal processing*, ACASP 92, pp.669-674, Grenoble, France.

Sjöberg J. (1995). *Nonlinear system identification with neural networks*, PhD Thesis, Dept. of Electrical Eng. Linköping University Sweden.

Slotine J.-J. E., Sanner R.M. (1993). Neural networks for adaptive control and recursive identification: a theoretical framework, In *Essays on control: Perspectives in the theory and its applications*, H.L. Trentelman, J.C. Willems (Eds.), Birkhäuser.

Sontag E.D. (1993). Neural networks for control, in *Essays on control: Perspectives in the theory and its applications*, H.L. Trentelman, J.C. Willems (Eds.), Birkhäuser.

Sorenson H.W., Alspach D.L. (1971). Recursive Bayesian estimation using Gaussian sums, *Automatica*, Vol.7, pp.465-479.

Steinbuch M., Terlouw J., Bosgra O., Smit S. (1992). Uncertainty modelling and structured singular-value computation applied to an electromechanical system, *IEE Proceedings-D*, Vol.139, No.3, pp.301-307.

Stoorvogel A.A. (1992). The discrete time $H_\infty$ control problem with measurement feedback, *SIAM Journal on Control and Optimization*, Vol.30, No.1, pp.182-202.

Styblinski M.A., Tang T.-S. (1990). Experiments in nonconvex optimization: stochastic approximation with function smoothing and simulated annealing, *Neural Networks*, Vol.3, pp.467-483.

Sussmann H.J. (1992). Uniqueness of the weights for minimal feedforward nets with a given input-output map, *Neural Networks*, Vol.5, pp.589-593.

Sutton R.S., Barto A., Williams R. (1992). Reinforcement learning is direct adaptive optimal control, *IEEE Control Systems*, April, pp.19-22.

Suykens J.A.K., Vandewalle J. (1991). Quasilinear approach to nonlinear systems and the design of $n$-double scroll ($n = 1, 2, 3, 4, ...$), *IEE Proceedings-G*, Vol. 138, No. 5, pp.595-603.

Suykens J.A.K., Vandewalle J. (1993a). Generation of $n$-double scrolls ($n$=1,2,3,4,...), *IEEE Transactions on Circuits and Systems-I (Special issue on chaos in nonlinear electronic circuits)*, Vol. 40, No. 11, pp.861-867.

Suykens J.A.K., De Moor B. (1993b). Nonlinear system identification using multilayer neural networks: some ideas for initial weights, number of hidden neurons and error criteria, *IFAC World Congress, Sydney Australia*, July 19-23, Vol. 3, pp.49-52.

Suykens J.A.K., De Moor B., Vandewalle J. (1993c). Stabilizing neural controllers: a case study for swinging up a double inverted pendulum, *NOLTA 93 International Symposium on Nonlinear Theory and its Applications, Honolulu, Hawaii*, Dec., Vol.2, pp.411-414.

Suykens J.A.K., Vandewalle J. (1993d). Between $n$-double sinks and $n$-double scrolls ($n = 1, 2, 3, 4, ...$), *NOLTA 93 International Symposium on Nonlinear Theory and its Applications, Honolulu, Hawaii*, pp.829-834.

Suykens J.A.K., De Moor B., Vandewalle J. (1993e). Neural network models as linear systems with bounded uncertainty, applicable to robust controller design, *NOLTA 93 International Symposium on Nonlinear Theory and its Applications, Honolulu, Hawaii*, Dec., Vol.2, pp.419-422.

Suykens J.A.K., De Moor B., Vandewalle J. (1994a). Static and dynamic stabilizing neural controllers, applicable to transition between equilibrium points, *Neural Networks*, Vol.7, No.5, pp.819-831.

Suykens J.A.K., De Moor B., Vandewalle J. (1994b). Nonlinear system identification using neural state space models, applicable to robust control design, *International Journal of Control*, Vol.62, No.1, pp.129-152, 1995.

Suykens J.A.K., De Moor B., Vandewalle J. (1994c). Stability criteria for neural control systems, *European Control Conference 95, Roma, Italy*, Vol.3, pp.2766-2771, Sept 1995.

Suykens J.A.K., De Moor B., Vandewalle J. (1994d). $NL_q$ theory and its extension to $\mu$ theory, K.U. Leuven, Department of Electrical Engineering, ESAT-SISTA, Technical report 94-63I.

Suykens J.A.K., Vandewalle J. (1994e). Generalized Cellular Neural Networks represented in the $NL_q$ framework, *IEEE International Symposium on Circuits and Systems (ISCAS 95), Seattle, Washington, USA*, pp.645-648, May 1995.

Suykens J.A.K., De Moor B., Vandewalle J. (1994f). Neural state space model framework for identification and control with global asymptotic stability criteria, *K.U. Leuven, Department of Electrical Engineering, ESAT-SISTA, Technical report 94-58I*, pp.75.

Suykens J.A.K., De Moor B., Vandewalle J. (1994g). $NL_q$ theory: a neural control framework with global asymptotic stability criteria, *K.U. Leuven, Department of Electrical Engineering, ESAT-SISTA, Technical report 94-83I*, submitted for publication.

Suykens J.A.K., De Moor B., Vandewalle J. (1994h). $NL_q$ Theory: Unifications in the Theory of Neural Networks, Systems and Control, *European Symposium on Artificial Neural Networks, Brussels Belgium*, pp.271-276, April, 1995.

Suykens J.A.K., Vandewalle J. (1994i). A Fokker-Planck Learning Machine for Global Optimization, K.U. Leuven, Department of Electrical Engineering, ESAT-SISTA, Technical report 94-57I, submitted for publication.

Suykens J.A.K., Vandewalle J. (1994j). Nonconvex optimization using a Fokker-Planck Learning Machine, *12th European Conference on Circuit Theory and Design, ECCTD '95, Istanbul, Turkey,* pp.983-986, August 1995.

Suykens J.A.K., Vandewalle J. (1995a). Discrete time Interconnected Cellular Neural Networks within $NL_q$ theory, *International Journal of Circuit Theory and Applications (Special Issue on Cellular Neural Networks),* Jan.-Feb, 1996.

Suykens J.A.K., Vandewalle J., Van Ginderdeuren J. (1995b). Feedback linearization of nonlinear distortion in electrodynamic loudspeakers, *Journal of the Audio Engineering Society,* Vol.43, No.9, pp.690-694.

Suykens J.A.K., Vandewalle J. (1995c). Learning a simple recurrent neural state space model to behave like Chua's double scroll, *IEEE Transactions on Circuits and Systems-I,* Vol.42, No.8, pp.499-502.

Suykens J.A.K., Vandewalle J. (1995d). $NL_q$ theory: a unifying framework for analysis, design and applications of complex nonlinear systems, *NDES'95-International Workshop on 'non-linear dynamics of electronic systems', Ireland Dublin,* pp.121-130, July, 1995.

Suykens J.A.K., Vandewalle J. (1995e). Global asymptotic stability criteria for multilayer recurrent neural networks with applications to modeling and control, *IEEE International Conference on Neural Networks (ICNN'95), Perth, Australia, 27 Nov - 1 Dec, 1995.*

Szu H. (1987). Nonconvex optimization by fast simulated annealing, *Proceedings of the IEEE,* Vol.75, No.11, pp.1538-1540.

The MathWorks Inc. (1994). Matlab, *Optimization Toolbox* (Version 4.1), *Robust Control Toolbox* (Version 4.1).

Thierens D., Suykens J., Vandewalle J., De Moor B. (1993). Genetic weight optimization of a feedforward neural network controller, *Artificial neural nets and genetic algorithms, Proceedings of the international conference in Innsbruck, Austria 1993,* (eds. R.F. Albrecht, C.R. Reeves, N.C. Steele), Springer-Verlag Wien New York.

Tolat V., Widrow B. (1988). An Adaptive 'broom balancer' with visual inputs, *IEEE International Conference on Neural Networks,* Vol.II, pp.641-647, San Diego, California.

Tollenaere T. (1990). SuperSAB: fast adaptive backpropagation with good scaling properties, *Neural Networks,* Vol.3, pp.561-573.

Tou J.T., Gonzalez R.C. (1974). *Pattern recognition principles*, Addison-Wesley.

Tsoi A.C., Back A.D. (1994). Locally recurrent globally feedforward networks: a critical review of architectures, *IEEE Transactions on Neural Networks*, Vol.5, No.2, pp.229-239.

Vandenberghe L., Boyd S. (1993). A primal-dual potential reduction method for problems involving matrix inequalities, *Mathematical Programming Series B, special issue on nondifferentiable and large scale optimization*, to appear.

van der Schaft A.J. (1992). $L_2$-gain analysis of nonlinear systems and nonlinear state feedback $\mathcal{H}_\infty$ control, *IEEE Transactions on Automatic Control*, Vol.AC-37, pp.770-784.

van der Schaft A.J. (1993). Nonlinear state space $\mathcal{H}_\infty$ control theory, in *Essays on Control: perspectives in the theory and its applications*, Eds. H.L. Trentelman, J.C. Willems, Birkhäuser.

van der Smagt P.P. (1994). Minimisation methods for training feedforward neural networks, *Neural Networks*, Vol.7, No.1, pp.1-11.

Van Overschee P., De Moor B. (1994). N4SID : Subspace Algorithms for the Identification of Combined Deterministic-Stochastic Systems, *Automatica*, Vol. 30, No. 1, pp.75-93.

Vidal P. (1969). *Non-linear sampled-data systems*, Gordon and Breach Science Publishers Inc.

Vidyasagar M. (1993). *Nonlinear systems analysis*, Prentice-Hall.

Wax N. (ed.) (1954). *Selected papers on noise and stochastic processes*, New York, Dover.

Weigend A.S., Gershenfeld N.A. (1993). Results of the time series prediction competition at the Santa Fe institute, *IEEE International Conference on Neural Networks*, San Francisco, CA, pp.1786-1793.

Weigend A.S., Rumelhart D.E., Huberman B.A. (1991). Generalization by weight-elimination with application to forecasting, *Advances in Neural Information Processing (3)*, R. Lippmann, J. Moody, D. Touretzky, Eds., pp.875-882.

Werbos P. (1974). *Beyond regression: new tools for prediction and analysis in the behavioral sciences*, PhD dissertation, Committee on Appl. Math., Harvard Univ., Cambridge, MA.

Werbos P. (1990). Backpropagation through time: what it does and how to do it, *Proceedings of the IEEE*, 78 (10), pp.1150-1560.

Widrow B. (1987). The original adaptive neural net broom-balancer, *Int. Symposium on Circuits and Systems*, pp.351-357, IEEE.

Widrow B., Rumelhart D.E., Lehr M.A. (1994). Neural networks: applications in industry, business and science, *Communications of the ACM*, March, Vol.7, No.3.

Wiener N. (1961). *Cybernetics*, The MIT Press and John Wiley & Sons, Inc. New York-London.

Wilkinson J.H. (1965). *The algebraic eigenvalue problem*, Oxford University Press.

Willems J.C. (1971a). The generation of Lyapunov functions for input-output stable systems, *SIAM Journal Control*, 9(1), pp.105-134.

Willems J.C. (1971b). Least squares stationary optimal control and the algebraic Riccati equation, *IEEE Transactions on Automatic Control*, AC-16(6), pp.621-634.

Willems J.C. (1972). Dissipative dynamical systems I: General theory. II: Linear systems with quadratic supply rates. *Archive for Rational Mechanics and Analysis*, 45, pp.321-343.

Williams R.J., Zipser D. (1989). A learning algorithm for continually running fully recurrent neural networks, *Neural Computation*, Vol.1, No.2, pp.270-280.

Wong E. (1971). *Stochastic processes in information and dynamical systems*, McGraw-Hill.

Yao Y., Freeman W.J. (1990). Model of biological pattern recognition with spatially chaotic dynamics, *Neural Networks*, Vol.3, pp.153-170.

Youssef H.M. (1993). Comparison of neural networks in nonlinear system modeling, *World Congress on Neural Networks '93*, Portland, Vol.IV, pp.5-9.

Zurada J.M. (1992). *Introduction to Artificial Neural Systems*, West Publishing Company.

# Index

## A

activation function, 19
approximation theorems, 23
ARMA model, 38
ARMAX model, 38
ARX model, 38
augmented plant, 120

## B

backpropagation, 29, 46
backpropagation through time, 32
BFGS formula, 56
bias term, 20
Box-Jenkins model, 38

## C

Chua's circuit, 182
clustering algorithm, 34
CNN, 168
condition number, 132, 137, 143, 150
confidence intervals, 59
conjugate gradient, 57
continuous simulated annealing, 196
convex hull, 63
correlation tests, 59
critic element, 4, 87
cross validation, 58
curse of dimensionality, 26

## D

DFP formula, 56
diagonal dominance, 128, 137, 143
diagonal scaling, 127, 136, 146, 148
direct adaptive control, 84
dissipative, 139
double inverted pendulum, 111
double scroll, 77, 162, 182
dynamic backpropagation, 31, 96, 153
dynamic programming, 86

## E

exogenous input, 120
extended Kalman filter, 48

## F

feedforward network, 21, 22
FIR model, 38
Fokker-Planck equation, 197

## G

Gauss-Newton method, 47
generalized delta rule, 29, 58
generalized gradient, 152
Gerschgorin's theorem, 129
global asymptotic stability, 127, 128, 132
gradient, 55
Green's function, 28

## H

$H_\infty$ control, 138
Hessian, 55
hidden layer, 20
Hopfield network, 22, 123

## I

identifiability, 43
indirect adaptive control, 84
innovations form, 41
interconnection matrix, 20
internal model control, 88
inverted pendulum, 104

## J

Jacobian, 31, 51, 97, 100, 183

## K

Kalman gain, 41, 49

## L

Lagrange multiplier sequence, 90
Levenberg-Marquardt, 55
LFT representation, 65
$L_2$-gain, 140
Linear Matrix Inequality, 147
LISP, 94
LQR, 99
LRGF, 172
Lur'e problem, 132
Lyapunov function, 127, 129

## M

mastering chaos, 162
McCulloch-Pitts neuron, 19
model predictive control, 88
model validation, 58
momentum term, 29, 57
multidimensional sampling theory, 202
multilayer perceptron, 19
$\mu$ theory, 145

## N

NARMAX model, 38
NARX model, 38
neural state space controller, 119
neural state space model, 41, 119
Newton method, 55
$NL_q$ system, 122

nonlinear least squares, 47, 57
nonlinear optimization, 55

## O

ordered derivative, 32
output error model, 38

## P

parallel model, 40
passive, 140
perturbed $NL_q$ system, 140
Pontryagin's maximum principle, 91
prediction error algorithm, 46, 52
pruning, 60

## Q

Q-learning, 85
quasi-Newton condition, 56
quasi-Newton method, 56

## R

radial basis function network, 22, 198
recurrent neural network, 31
recursive prediction error algorithm, 48
recursive stochastic algorithm, 196
regularization, 27, 60
reinforcement learning, 85

## S

sector condition, 132
sensitivity model, 31, 51, 96
Sequéntial Quadratic Programming, 105
series-parallel model, 40
sign-permutation equivalence, 43
spectral radius, 130
standard plant, 122
state space model, 38
state space upper bound test, 147
steepest descent, 55
storage function, 139
structured singular value, 145
supervised learning, 5
supply rate, 140
synthesis problem, 151
$\Sigma$ network, 24
$\Sigma\Pi$ network, 24

## T

two point boundary value problem, 90

## U

uncertainty, 145
unsupervised learning, 5

## W

weight decay, 60